TAKING FLIGHT

EDUCATION AND TRAINING
FOR AVIATION CAREERS

Janet S. Hansen and Clinton V. Oster, Jr., Editors

Committee on Education and Training for Civilian Aviation Careers

Commission on Behavioral and Social Sciences and Education
National Research Council

NATIONAL ACADEMY PRESS
Washington, DC 1997

NATIONAL ACADEMY PRESS • 2101 Constitution Avenue, NW • Washington, DC 20418

NOTICE: The project that is the subject of this report was approved by the Governing Board of the National Research Council, whose members are drawn from the councils of the National Academy of Sciences, the National Academy of Engineering, and the Institute of Medicine. The members of the committee responsible for the report were chosen for their special competences and with regard for appropriate balance.

This report has been reviewed by a group other than the authors according to procedures approved by a Report Review Committee consisting of members of the National Academy of Sciences, the National Academy of Engineering, and the Institute of Medicine.

This work is sponsored by the U.S. Department of Education and funded under contract no. EA9409001. The content of this publication does not necessarily reflect the views or policies of the U.S. Department of Education, nor does mention of trade names, commercial products, or organizations imply endorsement by the U.S. government.

Additional copies of this report are available from:
    National Academy Press
    2101 Constitution Avenue NW
    Box 285
    Washington, DC 20418
    800-624-6242 or 202-334-3313 (in the Washington metropolitan area). http://www.nap.edu

Contents

Acknowledgments

The Committee on Education and Training for Civilian Aviation Careers is deeply indebted to a number of individuals who made significant contributions to its work. First and foremost are the staff at the National Research Council. Gary Baldwin served as research associate throughout the project. He undertook special responsibility for tracking down frequently elusive literature and data; he also wrote invaluable background materials that assisted the committee in elucidating the issues, identifying the relevant evidence, and drawing appropriate conclusions. Senior Project Assistant Theresa Noonan coordinated the committee's meetings with great efficiency and prepared successive drafts of the manuscript. Alexandra Wigdor, as always, provided wisdom and guidance throughout. Christine McShane's editing added greatly to the clarity of the report. Elsa Riemer prepared the final version of the manuscript for publication with skill and care. Consultant Nancy Kober gave important assistance with writing in the late stages of the committee's deliberations.

The committee is also grateful for the willingness of literally dozens of individuals from government agencies, aviation trade and interest groups, colleges and schools, research organizations, industry, and the aviation press to help the committee accumulate essential information and evidence. A few of these individuals deserve special thanks for their good-natured and helpful responses to numerous requests throughout the study:

R. Gregg Carr, FAPA
Dr. Gary Eiff, Purdue University
Zee Ferris, Defense Manpower Data Center

Gary Kiteley, University Aviation Association
Dr. Larry Lackey, Federal Aviation Administration
Andrew Robinson, Federal Aviation Administration
Louis Smith, FAPA
Leslie Vipond, Federal Aviation Administration
Phillip Woodruff, Federal Aviation Administration

Clinton V. Oster, Chair
Janet S. Hansen, Study Director

TAKING FLIGHT

Executive Summary

Civilian aviation is a key part of the transportation system of the United States and a major contributor to national economic prosperity. As a result, the public and policy makers take a keen interest in its well-being. Although the commercial airlines were largely freed from government economic controls in 1978, federal agencies continue to have important oversight responsibilities for aviation safety, and the public still expects government to pay special attention to how the aviation industry is faring.

This study is one of several initiated by Congress in the years since deregulation to examine perceived problems in air transportation and in particular in the scheduled airlines. Its roots lie in concern about the continuing struggle of minorities and women to overcome barriers and prejudices that historically have restricted their access to many American workplaces. Like other industries, for many years aviation provided only limited opportunities for nonwhites and women. Explicit discrimination in hiring was accompanied by an internal culture that from the beginning of commercial aviation gave heavy emphasis to the masculine nature of flying. Even today, key airline jobs like pilots, maintenance technicians, and senior managers are still largely held by white men. This is frustrating for individuals from traditionally underrepresented groups who would like to fly or otherwise participate in this exciting field. It has also caused some to wonder whether the aviation industry will be able to find the workers it needs, with white men becoming a smaller and smaller part of the employed workforce and with the U.S. military, historically the source of many trained pilots and technicians, undergoing significant downsizing.

In light of these concerns, Congress directed the U.S. Department of Education to commission from the National Academy of Sciences a study on education and training for civilian aviation careers, mentioning in particular pilots, technicians, and managers. The department asked the committee conducting this study (1) to consider the implications of military downsizing for both training capacity and workforce diversity, (2) to review the capacity of civilian training institutions to meet future needs for aviation personnel, and (3) to focus specific attention on barriers facing minorities and women in gaining access to aviation careers and to examine options for attracting individuals from underrepresented groups into the aviation career pipeline.

The military is clearly going to be a less significant factor in producing civilian aviation personnel in the future. Highly skilled workers such as pilots and technicians will increasingly have to be trained by civilian institutions. The committee is confident that these institutions can handle the demand, but the way airline companies think about the preparation of their skilled workforce will need to change. Thanks both to the military and to a large general aviation sector encompassing both nonpassenger commercial and personal flying, airlines have been able to attract trained workers without playing much of a role in guiding or supporting the providers of that training. To guarantee the future availability of the workers they require, the airlines will need to work actively with schools and with government regulators to encourage the development of civilian programs offering standardized, airline-oriented training.

At the same time, the effort to open aviation jobs to a more representative cross-section of the population must continue. Formal policies that once barred minorities and women from many aviation jobs no longer exist, and progress has been made, but more remains to be done to ensure that opportunities are open to all. The challenge of diversifying the aviation workforce is a complex one, not least because severe labor shortages that would give employers strong economic incentives to identify and hire talented people from all races and both sexes have not characterized and may well not in the future characterize the commercial aviation industry. Nevertheless, the untapped potential of groups that have historically been underrepresented in aviation should be developed, for both ethical and economic reasons.

THE AVIATION INDUSTRY AND ITS WORKFORCE

Aviation companies, together with the many businesses associated with air transportation, generate millions of jobs for U.S. workers. Still, because air transportation is a capital-intensive enterprise, it employs relatively few people directly. Of the roughly three quarters of a million people in air transport, most work for the relative handful of companies that qualify as major airlines. Most airline workers are not pilots or technicians, two specialized occupations that the committee was specifically charged to study. In 1993, 101,000 people earned their living as pilots and 139,000 worked as aircraft engine mechanics.

Aviation makes an important contribution to the economy, but it is also strongly influenced by economic conditions. During recession, when the economy is growing slowly, airline passenger traffic growth usually falls off sharply and may even decline, as fewer people take vacations involving air travel and as businesses cut back on travel. Airlines respond to this decline by reducing employment, through slower hiring and employee furloughs. When the economy is doing well, airline travel usually grows quickly; airlines then call back furloughed workers and may also increase their hiring.

The industry's responsiveness to economic conditions is reflected in statistics on new hires, which gyrate substantially from year to year. New pilot hires over the past decade have ranged from a high of 13,401 in 1989 to a low of 3,256 in 1993. New technician hires reached 12,893 in 1989, but fell to 1,407 in 1993.

As these figures indicate, the late 1980s was a period of rapid airline expansion, which triggered a rare concern that demand for trained personnel would outstrip supply. The concern quickly faded as the industry underwent one of its frequent shifts in economic fortune. Its labor market went from perceived shortage to an oversupply of trained and available pilots and mechanics, who are increasingly called aviation maintenance technicians. The period highlighted, however, the fact that the issue of supply and demand in aviation is closely linked to the question of what constitutes a "qualified" job candidate. To fly or work on an airplane, individuals must meet minimum conditions laid down by the Federal Aviation Administration (FAA) and hold appropriate FAA certifications. These minima are generally far below the standards that the major airlines expect and, for the most part, have been able to demand, even in periods when labor markets were tight. Airlines face less an issue of whether "qualified" candidates are available than an issue of how much training and experience they can require of new hires and how much they will have to provide themselves.

A major reason people are concerned about making sure that the aviation industry has the employees it needs and that all individuals have equal access to jobs in this industry is that aviation jobs are widely perceived as good jobs: exciting, rewarding, and, perhaps most important of all, high paying. This perception stems in large part from the glamorous public image of the pilot, but it also has a basis in reality. Average earnings among employees in air transportation still exceed those of the typical American industrial worker, although not by as much as they did when aviation was protected from price competition. Pilots, particularly those working for major airlines, have always been among the nation's leading wage earners; senior captains today can make over $200,000 annually at some larger carriers. Pilot wages have also dropped in relative terms since deregulation; moreover, individuals who fly for smaller airlines or in other types of commercial aviation earn considerably less.

Aviation occupations, although changing, do not mirror the diversity of the overall American workforce. Although aviation employees as a group are not dramatically different in sex, race, and ethnic makeup from all employees, the representation of women and racial minorities varies substantially from occupation to occupation. Pilots and senior managers continue to be predominantly white and male; aviation maintenance technicians are less likely to be white than are pilots and managers but are mostly men. These employment patterns are in part the result of a history of explicit and implicit policies against hiring women and minorities for aviation jobs in the military and at the airlines, policies that have been the subject of legal challenge and government investigation for several decades. The aviation workforce is still affected by a history of discrimination. Although substantial progress has been made, concerns about discrimination still exist.

THE IMPACT OF MILITARY DOWNSIZING

Historically, the military has been an important source of trained professionals, especially pilots, for commercial aviation. The major carriers have relied on the military for about 75 percent of their pilots.

Since at least the late 1980s, the military services have been undergoing a fundamental reshaping, restructuring, and drawing down of force size. These efforts have led to reductions in the inventory of officers and enlisted personnel in aviation-related occupations in the services and have lowered the numbers of individuals being recruited into these specialties. In 1995, for example, the number of military personnel in their first year of service who were either pilots or pilot trainees was less than 40 percent of the total 10 years earlier. The number of first-year enlisted aviation maintenance personnel in 1995 was about 46 percent of the total in 1985.

Such reductions will eventually affect the ability of the civilian air carriers to draw on the military for trained aviation personnel. The major air carriers, faced with a reduced supply of military-trained pilots and technicians, will have to meet their future hiring needs by relying to a greater extent than they presently do on civilian sources of supply, as smaller air carriers already do.

As a major training ground for pilots and aviation maintenance technicians, the military's aviation-related workforce is not noticeably more diverse than the civilian aviation workforce. Minorities and women are better represented in military aviation specialties than they used to be, but—with the exception of minority male technicians—their presence in these jobs is small, still significantly lagging their representation in the overall population. Because the proportion of pilots who are minority and women in particular is very low, the military drawdown will not have much effect on the diversity of the pool of trained pilots available to the air carriers. It does mean, however, that opportunities are shrinking for minorities and women, as well as white men, to receive aviation training by joining the armed services.

CIVILIAN TRAINING FOR AVIATION CAREERS

In the future airlines will depend more heavily than in the past on civilian sources for initial training of their specialized workforce. A glimpse of what this future might be like came in the late 1980s, when rapidly rising demand for workers led to more hiring from civilian training programs. This situation drew attention to the differences between the structured and consistent backgrounds of individuals trained in the military and the diverse and more varied experiences of those trained through civilian aviation pathways. Before the issues surrounding civilian training could be fully explored, a reversal of economic fortune caused new airline hiring to plummet and interest to shift away from strengthening civilian training. Some promising new training initiatives were aborted or deemphasized.

These issues are emerging again, as airlines are now recovering economically and are again facing the pressures of expansion and the need to replace aging workforces. The question remains whether a training system less dependent on the military over the long run can be expected to provide the air transportation industry with the numbers and kinds of workers necessary to operate efficiently and safely. To answer this question the committee first identified five civilian pathways through which individuals can prepare for aviation careers.

Training Pathways for Specialized Aviation Occupations

The first pathway, military training, is discussed above. The other four pathways are in the civilian sector:

• Foreign hires. U.S. airlines have shown little inclination to hire pilots and aviation maintenance technicians who have trained abroad. If anything, the flow tends to be in the opposite direction, with American-trained pilots looking for jobs with foreign airlines when U.S. openings are scarce. Given the adequate supply of American-trained personnel, foreign hires are unlikely to be a significant source of personnel for U.S. airlines in the near future.

• On-the-job training. Pilots and technicians can currently earn the necessary FAA licenses and certification without attending a comprehensive formal aviation education program, by passing specific tests and fulfilling other requirements. Pilots can attend one of the more than 1,700 flight schools (600 of which are FAA-certified pilot schools), then work as flight instructors and in other commercial jobs that help them build flying experience and qualify for advanced certificates. This pathway is highly variable and idiosyncratic. Technicians can work under the guidance of certified technicians in aviation repair shops to qualify for FAA exams. The major airlines, however, seem to prefer to hire maintenance technicians who are graduates of FAA-certificated technical programs, most of which are found in technical schools and colleges.

• Collegiate training. Approximately 280 postsecondary institutions—including four-year and two-year colleges, vocational-technical schools, and other specialized schools—offer nonengineering aviation programs in such fields as flight education, maintenance, and airline management education. Almost 200 institutions are certificated by the FAA to offer maintenance training. The collegiate pathway is already the major training route for certified aviation maintenance technicians and is likely to become a more important training source for pilots. Unlike technicians, however, pilots are not ready for employment with the major airlines after completing undergraduate training. They typically must spend years working their way up through commercial, nonairline flying jobs before they have the qualifications the airlines require to secure an employment interview.

• Ab initio ("from the beginning") training. Under this pathway, airlines take carefully selected individuals with no flying experience and put them through intensive pilot training courses designed to meet the airline's specific needs. Ab initio training is popular with foreign airlines; Lufthansa and many other global carriers have ab initio training programs in this country. But U.S. airlines have not wanted or needed to pay for this type of "grow your own" training, because they have had access to a plentiful supply of candidates from the military and the nation's large general aviation sector.

Each of these pathways offers advantages and disadvantages to the airlines and prospective employees, and each produces (or potentially could produce) pilots or technicians with different characteristics. The military pathway for pilots, for example, costs industry very little, provides rigorous and demanding training that weeds out candidates who do not meet high standards, and is highly adaptable to technological change. It is therefore easy to understand why the military pathway for pilot training has been so attractive to civilian airlines as a source of new hires. But the military's role as a provider of trained personnel for the airlines is clearly on the wane. Foreign-trained personnel have not been much of a factor in U.S. aviation labor markets. U.S. airlines are unlikely to be willing to pay for the high costs of ab initio training as long as they have plenty of applicants from other sources.

That leaves on-the-job training and collegiate-based programs as the pathways that seem most likely to replace military training as the primary route to the major airlines. The committee expects that collegiate aviation increasingly will dominate on-the-job training because it has the potential to produce pilots and technicians specifically trained to standards recognized by commercial carriers.

Challenges for Civilian Training

Several challenges will have to be addressed to ensure that civilian training dominated by the collegiate pathway fulfills its potential to meet the specialized workforce needs of commercial aviation.

First, civilian programs must produce enough people to compensate for reductions in military personnel and supply airlines' demand as they expand and/or replace aging workforces. **The committee concludes that civilian training will be able to meet labor market demands, based on the demonstrated ability of the training sector to adapt to changing needs.** Moreover, the airlines are in a powerful position to influence the aviation labor market; at the extreme, they can do as many foreign airlines do and train their own workers. But nothing in the U.S. experience suggests that this will be necessary or likely. U.S. air carriers have many intermediate options for influencing the number and quality of the candidates available to them, short of undertaking and paying for training themselves. Many of these options were initially explored during the hiring boom of the late 1980s and could be expanded on now.

The committee further concludes that successfully exploiting these options will depend on the continuing professionalization and standardization of collegiate aviation programs. The airlines and the aviation industry as a whole ought to become more active and systematic partners in fostering the maturation of collegiate aviation and developing commonly recognized training standards to guide the development of appropriate curricula. Schools and industry together need to build the institutional mechanisms that foster standard-setting and recognize programs meeting the standards.

The committee recommends that collegiate aviation programs support the development of a system of accreditation similar to that found in engineering and business. The accreditation system should be developed in close cooperation with the airlines to ensure that curricula are responsive to their needs.

The committee further recommends that the commercial aviation industry support development of an accreditation system as well as provide more sustained and consistent support to individual aviation programs. Some companies already are active, but more widespread participation is needed.

The committee recommends that the FAA facilitate school-industry cooperation and the development of an aviation accreditation system. Industry-school relations are not yet well developed in aviation, and accreditation is in its early stages. The industry is accustomed to having the FAA involved in important aviation-related discussions, and its participation in efforts to build partnerships and develop an effective standard-setting mechanism will help legitimize these efforts.

The committee further recommends that the FAA review its training and certification requirements to ensure that they support rather than hinder the efficient and effective preparation of aviation personnel. The committee did not undertake a thorough review of FAA requirements that affect the training offered in collegiate aviation programs but did identify several concerns about the compatibility of FAA rules and airline-oriented training in collegiate institutions.

DIVERSIFYING THE AVIATION WORKFORCE

From the earliest days of flight, women and minorities shared the national fascination with airplanes. Unlike white men, however, women and blacks in particular faced a variety of roadblocks to participating in aviation, including legal barriers, segregation, and stereotypes that emphasized the masculine nature of flying. Aviation is not unique; its history reflects not only its own traditions but also broad societal patterns in America. Aviation, like the nation, has undergone change. Nevertheless, neither society as a whole nor aviation in particular is anywhere near being able to declare victory in the battle to provide equal opportunities and equal treatment to all individuals.

The committee's discussion of the difficult question of why so little progress has been made led to the conclusion that increasing the diversity of the aviation workforce (and especially broadening access to its highly skilled and most senior positions) is a task that must include but also extend beyond the industry itself. It must reach back long before the time that potential employees apply for jobs, because today very few women and blacks are certified to be pilots and mechanics. Enlarging the pool of people interested in and qualified for aviation careers can address two concerns simultaneously. It can increase the number of minorities and women available for employment. It can also forestall any future supply problems by ensuring that the nation's increasingly diverse workforce is being fully utilized by the aviation industry.

The committee concludes that the challenge of improving diversity in aviation must be addressed along three dimensions. Efforts must be made to develop the *interest* of individuals from underrepresented groups in undertaking aviation careers. There must be equal opportunities for minorities and women to develop the *basic academic competencies* to successfully pursue aviation careers if they choose. And any remaining *barriers* must be addressed that formally or informally have a disproportionate effect on the ability of minorities and/or women to pursue aviation careers if they have the interest and the basic academic competencies.

Developing Interest

Individuals from underrepresented groups need to know that aviation now offers career opportunities they can aspire to. Fortunately, many activities aimed at involving young people and others in aviation are already under way, sponsored by the federal and state governments, by private associations, by airlines and aircraft manufacturers, and by the many professional associations that represent companies or employees involved in all aspects of commercial and general aviation. Because of their historic exclusion from much of aviation, however, there is less of an aviation tradition among blacks and women than among white men, so voluntary programs are less apt to attract them without special recruitment efforts. **The committee recommends that all organizations seeking to encourage interest in and knowledge of aviation focus special attention on the continuing need to reach and involve individuals from groups who have been and still are underrepresented in the industry.**

Most aviation outreach activities focus on precollege age groups, thus missing the opportunity that exists for outreach and support in the collegiate institutions that enroll significant numbers of minority and women students in aviation education programs. These schools offer excellent opportunities for industry in particular to provide assistance to individuals to encourage their persistence in aviation and support for institutions that are demonstrating their ability to attract underrepresented groups into the aviation field. **The committee recommends that, to increase the pool of qualified applicants from underrepresented groups for pilot, aviation maintenance technician, and other positions in the aviation industry, airlines and other employers work aggressively to build linkages with the aviation programs at historically black colleges and universities and other schools and colleges with large minority and female enrollments.**

Resources of time and money are likely to be most efficiently used when the various groups involved in promoting aviation and aviation education work together. Many partnerships already exist, yet there is more room for progress. Cutbacks in the aviation education program at the FAA are a special concern, in particularly cutbacks affecting FAA publications that are widely used to disseminate basic information about aviation careers. **The committee recommends that industry work in partnership with state and private groups and the FAA to maintain basic aviation education and information services. The committee further recommends that the FAA and its parent agency, the Department of Transportation, reconsider their decision to cease providing (at no cost) basic information on the aviation industry and career opportunities that can be used by other aviation agencies and organizations to promote interest in the field.**

In small and large ways, aviation suffers from image problems that may hamper its attempt to diversify its workforce, from persistent use of the term *airmen* to describe pilots and technicians in FAA publications to narrow perceptions about aviation careers and the low regard with which technical jobs and vocationally oriented education and training are often held in this country. **The committee recommends that the responsible agencies and groups work to create more accurate public understanding of modern aviation careers and acceptance of the technical education needed to prepare for them.**

Basic Academic Competencies

Students need a solid grounding in mathematics and science to successfully pursue collegiate aviation programs. Research has provided a great deal of evidence about differences in mathematics and science participation and achievement among minorities and women at the precollege level. The task of improving mathematics and science competencies so that minorities and women can pursue aviation careers if they so choose is bound up in much larger questions of reforming precollege education and of improving the preparation of all students for science, math, engineering, and technology careers.

The committee recommends support for efforts to improve the general preparation of elementary and secondary school students in mathematics and science and stresses the continuing need to focus special attention on improving opportunities for and the academic achievement of minorities and women. The committee also recommends that those responsible for specialized aviation programs at the precollege level collaborate with larger systemic efforts to improve educational performance.

Barriers

Formal barriers restricting the employment of blacks and women in aviation may be gone, but there are still obstacles to overcome. One clear-cut one is the cost of training. More subjectively, the committee is under no delusions that invidious behavior has entirely ceased to plague the workplace. The more intangible barriers that result from such behavior, such as those that might affect how individuals are selected for jobs or the climate they encounter in schools and businesses, also need continuing attention.

The costs of specialized training for aviation careers, particularly flight training for pilots, are substantial. High costs are likely to pose a special barrier for students who come from households with below-average incomes, as many black and Hispanic individuals do. In addition to the expenses of flight training that would-be pilots face during their college years can be added the costs of flying time and certifications that individuals must accumulate after graduation—what we call "transitional training"—to qualify for airline interviews.

The committee recommends the establishment of financial assistance programs to help applicants for pilot positions meet the costs of flight and transitional training. Industry should take the lead in developing these programs.

Investigating barriers that may restrict access to aviation jobs for minorities and women leads to the issue of selection procedures, especially for professional pilots who are still predominantly white and male. The committee was not able to investigate this subject in depth but did discover that little information is available in the public domain about how pilots are selected for civilian employment. A more transparent process would be helpful both for training schools and for job candidates. **The committee recommends that airlines formalize and publicize their hiring criteria so that schools can develop appropriate programs of study and individuals can make informed decisions about training and career paths.**

Because the criteria used to select pilots are considered proprietary information by the airlines, most of the available research literature focuses on military pilot selection and training. The military's experience suggests that the predictive validities in its pilot selection system are low. Human factors researchers have highlighted the need for new job task analyses in aviation, reflecting the use of modern technology and the crew-oriented environment of the commercial airline cockpit. New instruments are being developed to measure traits not traditionally emphasized in pilot selection. Some preliminary evidence suggests that women and minorities may perform better on new measures than on some traditional instruments. **The committee recommends that all airlines examine their selection criteria and use procedures consistent with the best available knowledge of job tasks and effective crew performance.**

Finally, improving the diversity of the aviation workforce requires recognizing that the struggle of minorities and women to become full members of the aviation community is not yet over. The remaining job will in some ways require even more effort than what was needed to overcome blatant policies of discrimination and exclusion, because it means addressing habits or attitude and behavior that are much more difficult to identify and root out. Like other industries, aviation also has to shatter the so-called glass ceiling: the invisible, artificial barriers blocking women and minorities from advancing up the corporate ladder to management and executive level positions. **The committee recommends continuing efforts, vigorously led by top officials, to root out any remaining vestiges of discriminatory behavior in aviation training institutions and aviation businesses and to provide a favorable climate and truly equal opportunities for all individuals who wish to pursue careers in the aviation industry.**

1

Introduction

Civilian aviation is a key part of the transportation system of the United States and a major contributor to national economic prosperity. In 1993 the country's scheduled airlines transported 460 million people and carried 5.5 million tons of freight. In that same year, 6.9 million scheduled flights departed from U.S. airports carrying business people to meetings and pleasure travelers to vacation sites and visits with family and friends (Federal Aviation Administration, n.d.:Table 4-6). To make these flights possible, scheduled airlines employ over half a million people and work with aircraft manufacturers, airports, regulatory and safety agencies, and a host of other essential auxiliary industries that employ millions more. In the global marketplace of the late twentieth century, a healthy aviation industry is vital to a healthy economy.

As with other industries on which the nation's competitiveness hinges, the American public has a vested interest in the strength of the airlines. Aviation differs from most other industries, however, in that it has been extensively overseen and regulated by public agencies since its earliest days. For much of the history of the aviation industry, the federal government made the key decisions that determined route structures, fares, capacity constraints and other physical characteristics of the air transport system, work rules and worker qualifications, and more. Even though the era of tight federal economic controls ended with implementation of the Airline Deregulation Act of 1978, federal agencies continue to have important oversight responsibilities for aviation safety, and the public still tends to expect government to pay special attention to how the aviation industry is faring.

This study of education and training for civilian aviation careers might be viewed as one of several activities in recent years that carry on the legacy of government involvement. One such activity occurred in the late 1980s, when a period of robust commercial aviation activity coincided with the beginnings of downsizing in the military, which historically has served as a major supplier of trained pilots and maintenance technicians for private industry. Fears of a coming shortage of trained personnel led Congress to call for a study, which was carried out by a Pilot and Aviation Maintenance Technician Blue Ribbon Panel set up under the auspices of the Federal Aviation Administration (Blue Ribbon Panel, 1993). A subsequent airline "bust" in the early 1990s, brought on in part by recession and the Persian Gulf War, caused tremendous upheaval in the industry and resulted in the airlines' losing more money within a few years than they had earned in profits during the preceding half-century. These events spurred Congress to create a commission to investigate and make recommendations about the financial health and future competitiveness of the U.S. airline and aerospace industries (National Commission, 1993).

Another area of ongoing government concern has been the continuing struggle of minorities and women to overcome barriers and prejudices that historically have restricted their access to many American workplaces. Like other industries, for many years aviation provided only limited opportunities for nonwhites and women. But to the frustration of individuals from these groups who wanted to fly or otherwise participate in this exciting field, aviation appeared slow to open its doors even after formal barriers began to fall in the 1960s and 1970s. To individual frustration was added a more general concern about whether an industry like aviation, without making changes, could find the workers it needed in a country in which white men were becoming a smaller and smaller part of the employed workforce.

In light of these various concerns about the aviation workforce, Congress directed the U.S. Department of Education to commission from the National Academy of Sciences a study on education and training for civilian aviation careers. As the legislative history demonstrates, Congress was particularly interested in the issue of access to good jobs within the airline industry sector of civilian aviation. Special mention was made of pilots, aviation maintenance technicians, and managers.

The Department of Education set its charge to the committee in the context of military downsizing and of the traditional reliance of commercial airlines on specialized workers who have received their initial training outside the industry itself. The department asked the committee: (1) to consider the implications of military downsizing on both training capacity and workforce diversity, (2) to review the capacity of civilian training institutions to meet future needs for aviation personnel, and (3) to focus specific attention on barriers facing minorities and women in gaining access to aviation careers and to examine options for attracting individuals from underrepresented groups into the aviation career pipeline.

Given its charge and the time and resources available to it, the committee could not address all the variables related to employment in the aviation industry. A thorough examination of the role of labor unions in setting hiring standards, for example, or a review of Federal Aviation Administration certification and medical requirements for pilots, or analysis of industry fringe benefit policies that affect the relative attractiveness of aviation careers would require separate studies of their own. The committee's charge emphasized several specific careers most typically found in airlines; therefore, its analysis did not extend to conditions affecting the employment of air traffic controllers, aerospace engineers, or other career categories more characteristic of public or manufacturing employment in aviation than of air transportation.

On one level, the committee's review was straightforward. No one disputes that the military is going to be a less significant factor in producing civilian aviation personnel in the foreseeable future. Likewise, no one disputes that most of the highly skilled and senior jobs in aviation are still held by white men.

On another level, the issues facing the committee were significantly more complex. For one thing, as will become apparent time and again in this report, *quantifying* what is happening in the aviation workforce and its training paths is surprisingly difficult. Especially at the level of individual job categories like pilots and maintenance technicians, the aviation workforce is small enough that most standard employment surveys have large margins of error; although they can give an overall idea of employment levels, they are not very accurate as a gauge of year-to-year changes. Few statistics exist at all on managers, and none that separates senior management from the more junior ranks. Data collected by government agencies or by private groups frequently do not break down totals by sex or ethnic status. And as a journalist writing a series of articles on diversity in the industry recently reported, the airlines themselves are generally unwilling to provide data on the sex or ethnicity of their workforces (Henderson, 1995:34). Data on enrollments in aviation education programs are incomplete and capture only part of the civilian preparation system in any event.

Moreover, as a Smithsonian Institution history of women in aviation pointed out (in an observation equally applicable to minorities), "bare statistics are not very illuminating" in exploring the changing relationship between historically underrepresented groups and aviation (Douglas, 1991:107). Immense changes extending far beyond aviation—on issues such as desegregation and civil rights, equal employment opportunities, and the role of women in combat—have altered forever the society in which aviation operates and in which its younger and future workers have grown up. The first breakthroughs of women and black Americans into the most prestigious aviation jobs occurred comparatively recently. How much today's aviation workforce reflects continuing obstacles and discrimination and how much it reflects the time it takes individuals to work their way up through the system are questions with no precise, or at least scientific, answers. It is clear that underrepresented groups have made much progress. And it is clear that there is still more to be done.

This report reflects the fragmentary nature of the evidence the committee found on the matters in our charge. Much of our report focuses on pilots and maintenance technicians because the statistical data as well as qualitative analyses we found focus on these groups. This is not surprising. These are the two groups whose military training has been of the greatest benefit to the airlines. These are also the two groups within aviation who must be certificated by government authorities and for whom the most extensive education and training programs have grown up in the civilian sector. They are the only two airline occupations specifically included in the federal government's Standard Occupational Classification system, which determines the categories used to track employment in the decennial census and other studies. Nevertheless, even for these two occupational groups, data limitations restricted our ability to analyze some issues as thoroughly as we might have liked. Examples will be noted throughout the chapters that follow. Data limitations included imprecise statistics on airline hiring levels and on aviation education enrollments, the costs and quality of training, and airline selection and hiring practices.

When we turn our attention specifically to diversity and options to increase the representativeness of the aviation workforce, we view aviation careers more broadly. We directly address some barriers to the advancement of minorities and women in management, as well as examine what might be done to increase the number of female and minority pilots and maintenance technicians. We also point out reasons for encouraging everyone interested in promoting broader access to aviation to consider the full range of good jobs offered in the industry and not limit their vision to a handful of the most familiar occupations. When we can, we look at specific racial and ethnic groups, but much of our analysis focuses on blacks because much of the available information centers on them. Not as much is known about other groups (including religious groups as well as racial/ethnic minorities) or about the specific experiences of minority women, which may differ in important respects from the experiences of minority men. Here again, data limitations frequently interfere with our ability to paint as full a portrait as we would like of the circumstances of minorities and women in aviation. These limitations result chiefly from two things: the limitations inherent in public data collection efforts (which always involve difficult decisions about the tradeoff between the costs to both firms and public agencies of gathering additional data and the benefits resulting from the availability of more complete information) and the reluctance of firms (for competitive and other reasons) to make public information beyond what they are legally required to provide.

The report is organized as follows. Chapter 2 provides the context for the committee's analysis by first sketching the evolution of the aviation industry, then describing the key characteristics of the current industry and its workforce. Among the issues discussed in this chapter are the shaping role of government regulation and deregulation on the workforce, the factors that affect worker supply, demand, earnings, and qualifications, and the current gender and ethnic composition of the aviation workforce.

Chapter 3 examines the importance of the military as a source of trained professionals for civilian aviation. In this chapter the committee assesses the impact of military downsizing on the availability of minorities and women with aviation training and on training opportunities for these groups in the military.

Options for training the civilian aviation workforce are the focus of Chapter 4. Here the committee explores and compares the strengths and limitations of five training pathways, including military training, foreign hires, on-the-job training, collegiate training, and "ab initio" (from the beginning) training. The committee considers the challenges facing an aviation training system that will increasingly rely on civilian rather than military institutions.

Chapter 5 takes an in-depth look at the issue of diversity in the aviation workforce. The chapter analyzes past and present obstacles affecting the employment of women and minorities, examines ways to expand the pool of interested and qualified individuals, and recommends improvements in educational preparation, selection criteria, industry climate, and other areas that could help create a more diverse workforce.

2

The Aviation Industry and Its Workforce

THE AVIATION INDUSTRY

Orville and Wilbur Wright first flew an aircraft in 1903 at Kitty Hawk, North Carolina. That inaugural flight from a North Carolina dune spawned an enterprise that, by one estimate, now employs over 8 million people and annually contributes nearly three-quarters of a trillion dollars to the nation's gross domestic product (Wilbur Smith Associates, 1995).

This chapter gives an overview of the aviation industry and its workforce, in order to provide a context for our more detailed analysis of training issues in subsequent chapters. We review major developments in the history of civilian aviation, such as the waxing and waning of the federal regulatory role, that shaped the industry's organization and personnel practices. We describe the general structure of civilian aviation today and examine data on workforce size, wages, hiring, and composition, including available information on the diversity of aviation personnel. We look at licensing and certification policies for key careers and trends in worker supply and demand—all to help us understand better the pressure points that affect training and hiring practices within the industry.

The Evolution of the Industry

In January 1914, some 11 years after the Wright brothers' historic flight, the first known commercial air passenger operator in the United States was established in Florida (Komons, 1978:16). For three winter months, on a more or less regular basis, the St. Petersburg-Tampa Airboat Line shuttled passengers back and forth across 23 miles of water between the two cities after which it was named. Then, with the coming of spring, tourists moved north, and the airline went out of existence. The country's second known airline, Aero Limited, started in August 1919 to carry passengers between New York and Atlantic City, then moved to Florida and flew between Miami and Nassau. By early 1920, it, too, had ceased operations. Other small airlines came (and sometimes went) during the 1920s. In an era of remarkable invention and rapid change in transportation, commerce, and world affairs, these fledgling airlines rose and fell without action or interference from the government.

Government involvement in civilian aviation began, not with passenger airlines, but with the U.S. Post Office airmail service. (For a summary of the airline industry's early development, see Meyer and Oster, 1981: Chapter 2; greater detail can be found in Davies, 1972.) As early as 1916, funds for the carriage of airmail were provided from monies appropriated for "Steamboats or Other Power Boat Service," but not until 1918 did the Post Office translate its desire for an airmail service into action. Early airmail, like early passenger service, was not successful because of the relatively slow speed and the range limitations of early aircraft. Only with the advent of transcontinental airmail service were the real advantages of the air mode demonstrated. As an infrastructure for these transcontinental routes, the Post Office by 1925 had developed a system of landing fields and flashing beacons from New York to San Francisco capable of supporting both daytime and nighttime operations.

Thus, commercial air transportation in the United States began with a number of small passenger companies whose presence was often no more than transient and with a subsidized airmail service operated by the U.S. Post Office.

The first major piece of U.S. civil aeronautics legislation was the Contract Mail Act of 1925, known as the Kelly Act for its principal sponsor. Under the Kelly Act, the postmaster general was authorized to award airmail contracts to private airlines. Contracts were awarded through the issuance of route certificates, which gave a company the right to carry mail on a specific route. These route certificates initially had a 4-year limit, which was soon extended to a 10-year maximum limit. The Post Office worked to expand the domestic route system by avoiding competition on individual routes and using its power to award routes to streamline and rationalize the industry. Consequently, by 1933, the "Big Four"—United, American, TWA, and Eastern—collected nearly 94 percent of the $19.4 million paid to airmail contractors.

Because of some irregularities in the Post Office actions regarding airmail contracts, the successor to the Kelly Act, the Watres Act (the Air Mail Act of 1930), was nullified and existing airmail contracts were canceled for a brief period in 1934. The U.S. Army Air Corps took over carriage of the mail, serving fewer routes on a more limited basis. They proved to be ill equipped and untrained for even this more limited service and suffered a series of highly publicized accidents. Airmail service was quickly returned to civilian companies with temporary, competitive bid contracts. In June 1934, the Black-McKellar Act was passed, giving permanence to these temporary contracts by authorizing the postmaster general to extend their life for an additional nine months, after which they could be "continued in effect for an indefinite period" (Komons, 1978:266-267). On the shorter routes, new companies entered the field, but the "Big Four" kept the longer routes mainly because only they had the equipment and training to fly them.

Developments in transportation regulation during the 1930s were greatly influenced by the catastrophic economic scene. Against the backdrop of the National Industrial Recovery Act of 1933 and its interest in wage rules, the National Labor Board established a formula for pilot pay that continues to influence how pilots are compensated. In 1933 the board agreed to arbitrate an industry-wide pilot pay dispute. Pilots feared that the introduction of larger and faster planes would reduce their incomes if pay continued to be based on hourly wage rates. In its Decision 83, the Labor Board established a compromise pay formula based on seniority, hours flown, and the average speed of the plane. In 1934 the pilots' union successfully lobbied Congress to require airlines to comply with Decision 83 in order to hold an airmail contract (U.S. Department of Transportation, 1992:5-6).

The Great Depression in addition so affected American philosophy that the very basis of the U.S. economic system, the concept of open competition, came under attack. Much economic regulation in the 1930s was a response to what was considered to be excessive or cutthroat competition. Additional reasons for regulation can arise, moreover, when an inherently competitive industry receives subsidies; under such conditions, bidding for new businesses can be at prices well below cost, allowing for expansion with the use of subsidy to make up the cost/ price difference. Such behavior did occur among the airlines in the 1930s, and the government sought to subdue competition of this sort so as to minimize subsidy payments.

Developed in this era, the Civil Aeronautics Act of 1938 formed the basis for federal aviation policies and authority and remained the primary influence until the time of deregulation in 1978. The 1938 act created the Civil Aeronautics Authority, which was reorganized into the Civil Aeronautics Board (CAB) in 1940. Among the Civil Aeronautics Authority's first actions under the new law was to grant "grandfather" rights, which gave existing airlines permanent certificates for all of their existing route authorities, as long as they met regulatory requirements. These certificates put a stamp of approval on the airline structure that the Post Office had developed, essentially freezing the system of the Big Four plus 12 independents. Through mergers in the intervening years, these 16 carriers evolved into the 10 domestic "trunk" or primary airlines that existed when deregulation started.[1] Although there were many applications to enter into

---

[1] As of October 1978, these airlines were American, Braniff, Continental, Delta, Eastern, National, Northwest, Trans World, United, and Western. In addition, Pan American was a trunk airline for international routes.

scheduled domestic airline service, no new trunk carriers were allowed to enter the industry under CAB regulation.

A basic tool of CAB regulation was its authority to issue "Certificates of Public Convenience and Necessity." Each certificate applied to a specific route, much like the certificates issued by the Post Office. An airline was required to have such a certificate before it offered scheduled airline service. The only exceptions were commuter airline operations with small aircraft.

Even though no new trunk carriers were permitted, the CAB, with considerable prodding from Congress, did allow some new entrants into the industry under special restrictions. In 1944 the CAB initiated an experiment to expand air service to small communities. This experiment eventually led to the formation of the so-called local service airlines,[2] intended to provide feeder service for the trunk airlines. The certificates of the local service airlines were changed from experimental to permanent in 1955. Through a combination of growth, government subsidy, and CAB route policies, many of these local service airlines came to resemble small trunk airlines in their operation. Even so, they retained a distinct regional focus in their route networks prior to deregulation.

A second type of entrant into the industry was the supplemental carrier. These carriers were permitted to offer only nonscheduled charter flights to groups of travelers. Initially, supplemental carriers were established by the CAB on an interim basis, but in 1966 they were given permanent status. Many of the applications to become scheduled carriers came from this group of supplemental carriers. Conversely, most of the trunk airlines conducted some charter operations.

The authority to issue certificates also allowed the CAB to control the route structure of each airline. Each certificate applied to a single route and specified the cities on either end of the route and the intermediate stops. The specification of intermediate stops was quite detailed and often included stops that had to be made, stops that could be made, and stops that could not be made. In addition, there were occasionally restrictions on carrying passengers between intermediate stops. Through the use of these certificates, the CAB could control the route structure of each airline as well as the numbers of carriers permitted to compete in each city-pair market.

The CAB used route awards to pursue diverse policy objectives. Monopoly route awards were often given the applicant in the weakest financial position in an attempt to strengthen that carrier and maintain stability in the industry. Route awards were also used as a way to reduce the subsidies paid to local service airlines for service to small communities. Thus, local service airlines were sometimes given longer routes, which were believed to be more profitable, with the intention that these profits would cross-subsidize the losses on low-density short-haul routes.

---

[2] As of October 1978, these airlines were USAir (formerly Allegheny), Frontier, Hughes Airwest, North Central, Ozark, Piedmont, Southern, and Texas International.

The certificates were also used to control exit, since a carrier could not stop serving a route without CAB approval. Some exit occurred, though, as the trunk airlines were permitted to turn over a few of their low-density short-haul routes to local service airlines. Starting in the late 1960s, the local service airlines and to a lesser extent the trunk airlines were permitted to hand over a few of their routes to noncertificated commuter airlines (who flew regular schedules but used smaller airplanes than the trunk and local service airlines).

The CAB also controlled airline fares. It could regulate fares in two different ways: (1) by approving, modifying, or rejecting requests for fare changes filed by individual carriers and (2) by directly setting either the exact fares or a narrow range of permissible fares. When the CAB set fares directly, it used a formula that had distance as its only variable. Thus, for the most part, all routes of the same length had the same fares for all the airlines.

The CAB was also clearly able to control the structure of the industry. In addition to influencing the size of the firms through the awarding of new routes, it had the authority to approve mergers of airline companies. In general, the CAB took a restrictive view of mergers, granting approval only to maintain industry stability when one of the merging airlines was in serious financial trouble.

The extent of these controls and the manner in which the CAB applied them had profound impacts on the development of the industry. Some of these impacts are still seen today. Under CAB regulation, for example, the airlines had very little incentive to keep labor costs, and particularly pilot salaries, low. If an airline involved in contract negotiations with pilots pushed for pilot salaries below the industry average, it risked a strike by the pilots. During that strike, the airline would lose considerable revenue from canceled operations. If the airline ultimately prevailed and the pilots accepted salaries below the industry average, what would the airline gain? It could not increase its market share by lowering its fares because fares were regulated by the CAB. Similarly, it could not increase its market share by expanding into new markets, because the CAB controlled routes and entry and seldom allowed carriers to enter new markets, particularly those already served by an incumbent carrier. With lower costs, it could earn more profits on each flight, but too much financial success would result in other airlines gaining preference for new route awards. Thus the gains from lower labor costs were small and rarely were judged by airline management to outweigh the losses incurred during a strike. The airlines knew that the CAB routinely adjusted fares to follow changes in the average industry costs. Thus, there was an incentive for airlines to grant wage increases as long as they were in line with average industry wage practices, because the resulting cost increases would be translated into higher fares for all airlines.

Working in concert with this reduced incentive to keep wages low was extremely high growth in airlines' productivity during virtually the entire post-World War II period, as can be seen in Table 2-1. Between 1950 and 1980, enplanements (the number of passengers boarding aircraft) increased by a factor

TABLE 2-1   U.S. Scheduled Airline Enplanements, Passenger Miles, Revenues, and Employees, Selected Years

Year	Enplanements (thousands)	Revenue Passenger Miles (millions)	Passenger Revenue (millions of 1993 dollars)	Employees
1950	19,220	10,243	$3,646	86,057
1960	57,872	38,863	11,657	167,603
1970	169,922	131,710	28,404	297,374
1980	296,903	255,192	49,187	360,517
1990	465,560	457,926	64,605	545,809

NOTE: Data before 1970 are taken from Civil Aeronautics Board, *Handbook of Airline Statistics, 1973 Edition* (March 1974). Data from 1970 on are from Air Transport Association, *Air Transport: The Annual Report of the U.S. Scheduled Airline Industry* (various years).

SOURCE: Morrison and Winston (1995:7). Reprinted by permission.

of 15, whereas employment increased only by a factor of 4. Most of the productivity growth came from improved aircraft technology. Aircraft became larger and faster and able to fly greater distances without stopping. During this period, revenue passenger miles increased more rapidly than enplanements, indicating that the average trip taken by a passenger became longer. The pace of technology growth accelerated with the spread of jet aircraft in the 1960s and of wide-body jet aircraft in the 1970s. Pilots took the view that, since they were able to produce more passenger miles with the newer aircraft, they should be paid more for their greater productivity. These productivity gains, coupled with the lack of regulatory incentive to keep wages down, resulted in pilots, particularly the more senior pilots, becoming highly paid.

Spurred by "a considerable body of academic research that was critical of regulation," the regulation of the airline industry was gradually relaxed beginning in 1975 and the Airline Deregulation Act was enacted in October 1978 (Transportation Research Board, 1991:28-30). Between 1975 and 1978 the CAB encouraged competition by opening certain routes to additional carriers and by giving airlines more discretion in pricing discount seats. The 1978 Deregulation Act ended (over time) CAB's authority over routes and domestic fares and dissolved the agency itself in 1985.

The changes to the industry since deregulation have been dramatic. As with any major public policy change, there have been winners and losers. In general, average air fares are lower than they were likely to be under continued regulation, more cities are receiving more service, and the industry has continued to improve its productivity, even in the absence of the dramatic technological progress that drove productivity improvement in the past. These factors have helped fuel a significant increase in airline travel over the past 20 years, with some fluctuations depending on overall economic conditions.

Deregulation has had important impacts on employment in the industry. The increase in airline travel increased the need for airline employees. However, this rise in employment was not evenly spread throughout the industry. Escalating competition in the wake of deregulation has meant growth for some airlines and the creation of some entirely new airline companies. It has meant that airlines previously constrained by regulation to intrastate markets, such as Southwest, could expand across state lines. It also has meant shrinkage of other airlines and the bankruptcy of some previously well-established carriers like Pan American, Eastern, and Braniff. Others of the pre-deregulation airlines were absorbed through merger.

While the large commercial airline portion of the industry is the main focus of the committee's attention, it is only a portion of the U.S. civilian aviation industry. For the purposes of this study, it is important to place commercial airlines in the larger context of civilian aviation because other parts of the civilian aviation industry play an important role as a training ground for pilots, maintenance technicians, and other airline personnel.

A Classification System for Civilian Aviation

An important foundation for the committee's work is a clear and consistent way of describing the various parts of civilian aviation. Unfortunately, classifications and terminology vary from source to source. For example, which carriers are and are not included in descriptions and statistics on airlines or "air carriers" are not always the same. Even within a single federal agency, the Federal Aviation Administration (FAA), definitions are not always consistent among agency publications (compare, for example, the glossaries in Federal Aviation Administration, n.d.(b) and 1995); the categories used in statistical tables vary depending on what is being described. For example, the *FAA Statistical Handbook of Aviation* (Federal Aviation Administration, n.d.(b)) uses different groupings of air carriers depending on whether it is reporting on airport activity, the civil air carrier fleet, operating data, general aviation aircraft, or aircraft accidents or whether it is defining terms in its glossary. This seems to be at least partially the result of different reporting requirements applied at different times to different kinds of air carriers by the various federal agencies that have had jurisdiction over aspects of passenger and cargo operations at some point in American aviation history. It may also be a function of rapid changes in the industry following deregulation in 1978. We try to be clear in this report about exactly what data we are describing, and we urge readers to be cautious when comparing the numbers presented in this report to those found elsewhere.

Nonmilitary aviation in the United States can be generally divided into two broad categories: commercial aviation and general aviation (Figure 2-1).

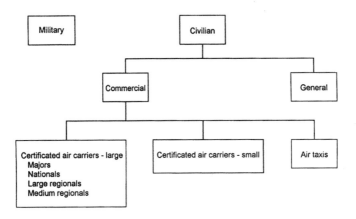

FIGURE 2-1   Aviation categories.

Commercial Aviation

Commercial aviation generally refers to businesses that carry passengers or cargo for hire or compensation. The FAA divides commercial air carriers into two broad categories: certificated air carriers and air taxis. The key difference between the categories involves aircraft size as measured by the number of seats or the maximum permissible payload of the aircraft flown. Below we follow the definitions used by the FAA in reporting on operating data for U.S. air carriers (Federal Aviation Administration, n.d.(b):Chapter 6).

Certificated Air Carriers These carriers fly planes of 30 or more seats and payload capacities of 7,500 or more pounds. Their operations are regulated under Part 121 of the Federal Air Regulations (contained in the Code of Federal Regulations: 14 CFR Parts 1 through 199). So-called large certificated air carriers operate aircraft with 60 or more seats or a maximum payload of 18,000 pounds or more. In some FAA statistical reports, these large certificated air carriers are further grouped into majors, nationals, large regionals, and medium regionals, according to their total operating revenues. The airlines in each group therefore change as carrier revenues change; Table 2-2 indicates the carriers that were in each group in 1992 and 1993. Some FAA statistics on large certificated air carriers include all four groups; others omit the medium regionals and report only on the other three groups. "Small certificated air carriers" fall between the large certificated carriers and air taxis (see below). While they operate under Part 121, they are often grouped with the scheduled air taxis in FAA statistical reports (e.g., Federal Aviation Administration, 1995:K-3) and, along with the scheduled air taxis, are commonly called "commuter air carriers."

TABLE 2-2   U.S. Large Certificated Air Carriers, 1992 and 1993

Major Carriers	National Carriers	Large Regional Carriers	Medium Regional Carriers
America West	Air Wisconsin	Air Transport Intl.	Aerial
American	Air Wisconsin Corp.	American Intl.	Airmark
Continental	Alaska	Amerijet	Atlas Air
Delta	Aloha	Arrow	AV Atlantic
Federal Express	American Trans Air	Braniff International	Buffalo
Northwest	Atlantic Southeast	Carnival	Casino Express
Southwest	Business Express	Challenge Air Cargo	Continental Micronesia
Trans World	DHL Airways	Executive Airlines	Eagle Airlines
United	Emery	Express One	Empire
United Parcel	Evergreen	Florida West	Fine Airlines
USAir	Hawaiian	Key	Great Americans
	Horizon Air	Kiwi	Intl. Cargo Express
	Markair	MGM Grand	Jet Fleet
	Midwest Express	Morris	Miami Air
	Southern Air	Northern Air	Million
	Sun Country	Private Jet	North American
	Tower	Reeve	Patriot
	USAir Shuttle	Reno	Ryan Intl.
	Westair	Rich	Sierra Pacific
	World	Simmons	Spirit Air
		Trans Continental	Trans Air Link
		Trans States	Trans American Charter
		UFS, Inc.	Ultrair
		Zantop	Wilbur's
			Worldwide
			Wrangler

SOURCE: Federal Aviation Administration (n.d.(b):Table 6.1).

Air Taxis Air taxis operate under Part 135 of the Federal Air Regulations and fly aircraft with fewer than 30 seats or payload capacities of less than 7,500 pounds. Scheduled air taxis perform at least five round trips per week between two or more points and publish flight schedules for these flights. In 1992 and 1993 there were 141 small certificated and scheduled air taxis combined (Federal Aviation Administration, n.d.(b):6-12). In addition to these scheduled commuters, 3,764 aircraft were primarily used in 1993 for what is called on-demand air taxi service (p. 8-3).

Affiliations and Feeder Relationships Among Airlines Following deregulation, the major air carriers have pursued various arrangements with smaller air carriers to ensure that passengers will be fed from small communities to hub airports where they can connect to jet flights. "Feeder" carriers typically operate as extensions of the major carriers under such names as American Eagle, the Delta Connection, Northwest Airlink, and United Express. These commuter operations have code-sharing or joint marketing arrangements with the majors that allow them to share the major carrier's two-letter designator code in reservation systems and published flight guides. They also coordinate flight schedules and baggage handling. Sometimes the major carriers hold equity interests in the smaller companies, and in a few cases they own them outright.

General Aviation

The FAA defines general aviation as encompassing all facets of civilian aviation except commercial air carriers. Although general aviation is not a focal point of our study, it plays an important role in the aviation training system in the United States and also provides employment for individuals with specialized aviation skills. Planes are used in general aviation for a variety of purposes, some of which are clearly commercial, even though they do not involve carrying passengers. General aviation includes firefighting, agricultural crop dusting and seeding, reforestation, insect control, aerial observation, business transportation (not for compensation), instructional flying, personal flying, glider towing, and parachuting. The FAA (n.d.(b):8-3) estimates that there were about 176,000 general aviation aircraft in use in 1993 (including 3,764 general aviation aircraft being used primarily for nonscheduled air taxi service). These ranged from corporate multiengine jets to amateur-built single-engine piston planes, balloons, and dirigibles. Technical services for general aviation (maintenance, repairs, and flying lessons, for example) are provided by hundreds of so-called fixed-base operators, who can be found at both major and small airports in the United States. There were 13,228 airports in 1993 (Federal Aviation Administration, n.d.(b): Table 3.1).

Aviation and Intermodal Transportation

The airlines do not operate in isolation but are part of an intermodal transportation system. The most sophisticated airline system in the world is of limited value if either customers or employees cannot access airports easily through a variety of ground transportation choices. Airport ground access has been a key focus for urban transportation planners since the 1950s and 1960s, when air travel was recognized as a significant factor in economic growth. Three decades ago, however, airport access meant creating highway systems that moved people, and to a lesser extent goods, to the airport. Highways provided the most simple solution to what was then a simple problem.

Over the last two decades, rapid growth of air transportation at large airports has created the need for improved integration of air transportation with multimodal ground transportation. Today, access has become much more complex, and planners must contend with a host of ancillary factors such as air quality, noise pollution, congestion, land use, wildlife and wetland protection, and others. Ground transportation at large airports is dominated by privately owned automobiles, a situation that creates a variety of problems (for a more complete discussion see Nettey, 1995). At the airport itself, there are problems of increased congestion of vehicular traffic both coming and going within the airport terminal complex. Airport congestion problems are often compounded by the obstructed flow of vehicular traffic by cars in search of increasingly scarce parking in proximity to the terminal. Both of these problems are worsened by the increase in demand for transportation during peak travel periods, when transportation delays are least tolerable and most economically wasteful.

Moreover, access no longer means moving just people to and from the airport during the most popular travel hours. The growth of air cargo, particularly the overnight package express services such as Federal Express and United Parcel Service, places additional demands on timely and predictable truck linkages with airports. These services rely on extensive truck pickup and delivery systems that must be able to meet rigorous time schedules both for getting packages to the airport in time for their flights to cargo hub airports and for delivering packages to the customers by the promised time. Easy ground access to airports is critical to the success of these services.

Aviation and the Economy

Airlines and their associated businesses play an important and highly visible role in the economy. Because air transportation is capital intensive, however, it employs relatively few people. Employment in air transportation and aircraft manufacturing together accounted for barely 1 percent of U.S. nonfarm employment in 1993 (Table 2-3). Air transportation itself (including airport operations) employed slightly less than 0.7 percent of nonfarm workers. Aviation, however, is highly visible and its contributions to the U.S. economy far exceed its direct employment.

Since 1989, with sponsorship from the aviation industry,[3] Wilbur Smith Associates has published a series of reports attempting to assess civilian aviation's overall contribution to the U.S. economy (Wilbur Smith Associates, 1995). These reports look at commercial and general aviation as well as airport operations and aircraft manufacturing and consider direct impacts, indirect impacts, and induced impacts. Direct impacts are financial transactions that occur due to the provision of air services and include expenditures by airlines, airport tenants, air cargo firms, aircraft manufacturing, fixed-base operators, flight schools, asso-

---

[3]The Partnership for Improved Air Travel, an industry group, commissioned reports in 1988 and 1990; Lockheed Martin asked Wilbur Smith Associates for updates in 1993 and 1995.

TABLE 2-3   Employment by Major Industry and Manufacturing Group, 1993
(numbers in thousands)

Total nonfarm employment	110,525
Goods producing (includes manufacturing)	23,256
Service producing	
Transportation, communications, and public utilities	5,787
Trade—retail, wholesale	25,675
Finance, real estate, insurance, services	36,990
Government	18,817
Employment in selected categories: manufacturing and transportation	
Total manufacturing	18,003
Aircraft and parts	542
All other manufacturing	17,461
Total transportation and transportation services	3,587
Air transportation	737
Railroad transportation	250
Local/interurban passenger transit	374
Trucking/warehousing	1,685
Water transportation	167
Pipelines, except natural gas	18
Transportation services	356

NOTE: Data are based on surveys of industry establishments; annual averages.

SOURCE: Bureau of Labor Statistics (1995a:Table 48).

ciated ground transport firms, and so forth. Indirect impacts are financial transactions that occur due to the use of aviation, including expenditures by visitors who arrive by air, travel agents, expenditures by business aviation, and so forth. Finally, induced impacts account for the ripple effect of economic activity, whereby people who earn money through the direct and indirect effects spend that money on supplies and on such consumer goods as food, clothing, and housing. By using input-output analysis and multipliers, these induced or ripple effects can be estimated. The techniques used in these studies are standard and well-accepted techniques for estimating economic impacts, although, as with all studies using these techniques, there is considerable uncertainty about the estimates of impacts, particularly the induced impacts.

According to the most recent update, in 1993 the annual economic impact of commercial aviation was estimated at almost $725 billion. About 8.3 million jobs, with wages of $215 billion, can be attributed to commercial aviation. There were also some much smaller benefits for general aviation. Thus, whereas direct air transportation employment may be responsible for less than 1 percent of jobs, commercial aviation accounts for over 10 percent of U.S. gross domestic product. The benefits of aviation are widely dispersed throughout the United States, and substantial aviation-related employment is found in all of the 50 states. Aviation also makes an important contribution to the U.S. balance of trade. Aircraft manufacturing has traditionally been a major segment of U.S. exports. Airline operations are also an important contributor to the balance of trade because of, among other things, the large number of foreign travelers using U.S. airlines for both international and domestic travel. Aviation is also a highly visible industry that often garners more attention than its contribution to the economy might suggest. Air safety, for example, is intensely scrutinized by the media and the public, even though deaths in air travel are few compared with other transportation industries.

Aviation makes an important contribution to the economy, but it is also strongly influenced by economic conditions. Both before and following deregulation, commercial passenger aviation has been very sensitive to economic conditions. During recession, when the economy is growing slowly, airline passenger traffic growth usually falls off sharply and may even decline, as fewer people take vacations involving air travel and as businesses cut back on travel. As travel declines, the airlines respond by reducing employment, through slower hiring and employee furloughs. When the economy is doing well, airline travel usually grows quickly; the airlines then usually call back furloughed workers and may also increase their hiring.

THE AVIATION WORKFORCE

Employment, New Hires, and Wages

Employment

Most people involved in air transport work for the relative handful of companies that qualify as major airlines. Most airline workers are not pilots or technicians, two specialized occupations that the committee was specifically charged to study.

In 1993, 737,000 people worked in the air transport industry (Table 2-4). Another 542,000 people were involved in manufacturing aircraft and aircraft parts, and 53,000 people worked for the Federal Aviation Administration overseeing, regulating, and promoting the nation's airways and aviation system.

These numbers reflect employment as measured by federal surveys of business establishments, classified by industry groups. Another, though not strictly comparable, source of national data on aviation employment are household surveys, in which respondents are classified by their primary occupation. These surveys utilize standard occupational codes developed by the U.S. Department of Commerce, which recognize three specialized aviation-related transportation occupations: airline pilots,[4] aircraft mechanics,[5] and aerospace engineers. Pilots

---

[4]The occupational classification system used by federal agencies conducting household surveys

TABLE 2-4   Aviation-Related Employment, 1993 (in thousands)

Employment in key aviation-related industries	
Air transport	737
Manufacturing: aircraft and parts	542
Government: FAA	53
Employment in specialized transportation occupations	
Airplane pilots	101
Aircraft engine mechanics	139
Aerospace engineers	83

NOTE: Data on employment in the air transport and aircraft manufacturing industries are based on surveys of industry establishments. Data on employment in specialized transportation occupations are based on household surveys and are not directly comparable with the establishment data.

SOURCE: Bureau of Labor Statistics (1994:Table 22; 1995a:Table 48); Federal Aviation Administration (n.d.(b):Table 1.2).

and technicians are engaged heavily although not exclusively in air transport; aerospace engineers are more likely to work in aircraft manufacturing. In 1993 (Table 2-4), there were 101,000 people who earned their living as pilots, 139,000 who worked as aircraft engine mechanics, and 83,000 who were employed as aerospace engineers (mostly in manufacturing rather than air transport).[6]

As best we can determine from comparing information from different sources, the airlines employ between half and two-thirds of the individuals employed as pilots and a smaller proportion of aircraft mechanics. The airlines employ perhaps three-quarters of the individuals who report themselves as holding air transportation-related jobs. Table 2-5 summarizes employment at the airlines large enough to be required to file Form 41, Schedule P-10, with the U.S. Department of Transportation. In 1993, the large certificated air carriers (excluding the medium regionals) reported a weighted average of 549,784 full-time

---

actually refers to "pilots and navigators." In this report we refer to the category as "pilots" since "navigator" is now an obsolete term.

[5] Avionics technicians are not included in the category "aircraft mechanic" but instead are counted in the occupational category of electronic repairers, communications and industrial equipment (Bureau of the Census, 1980:O-67). Avionics technicians are not separately identified in this category. However, since these individuals may also have mechanics' certification, they may already be accounted for in the mechanic totals.

[6] The figures in Table 2-4 should be interpreted as approximations, since they are based on survey data; and the small size of the categories involved means that standard errors are comparatively large. Moreover, the table understates the number of aviation maintenance personnel because the underlying data source reports only aircraft engine mechanics and not aircraft mechanics other than engine. Data from the 1990 decennial census indicate that there were about 32,000 aircraft mechanics other than engine.

TABLE 2-5 Weighted Average Number of Full-Time Employees, by Labor Category, Large Certificated Air Carriers, 1993

Carrier Type	Pilots	Other Flight	Passenger and General Services/ Administration	Maintenance Labor	Aircraft and Traffic Handling	Passenger Handling	Cargo Handling	Miscellaneous	Total
Majors	46,006	3,014	80,179	52,977	42,845	70,249	49,691	158,953	503,914
Nationals	4,333	922	4,343	3,510	4,270	4,145	1,471	8,960	31,954
Large regionals	3,002	1,310	1,734	2,234	1,463	841	364	2,968	13,916
Total	53,341	5,246	86,256	58,721	48,578	75,235	51,526	170,881	549,784

NOTE: All large certificated air carriers, which include all major and national airlines and those regional carriers with operating revenues over $20 million, are required to file Form 41, P-10 with the U.S. Department of Transportation annually. The number of employees reported reflects the weighted average number of full-time employees who received pay for any part of the calendar year. Data for employees by labor category are allocated on a basis consistent with that used in the allocation of salaries from Form 41 for financial reporting purposes. Data for "miscellaneous" category include general management, aircraft control personnel, trainees and instructors, record keeping, traffic solicitors, other personnel and transport-related. Data not available for Westair (national) and Key (large regional). Data reflect numbers provided by the airlines for each occupational category. Totals have been adjusted to reflect the sum of the numbers provided by the airlines when the airline input contained a computational error.

SOURCE: RSPA/DOT Form 41, Schedule P-10, 1993 on file at the U.S. Department of Transportation.

employees, of whom 53,341 were pilots and 58,721 were maintenance laborers, most but not all of whom were probably mechanics.[7] (Except for a small number of general managers, most management personnel are distributed across other occupational groupings and cannot be separately identified. There are also no comparable data for the medium regional group of large certificated air carriers, or for commuters and nonscheduled air taxis.)

The employment data reported by the airlines reflect those persons employed directly by the airlines and as such may understate the number of employed persons who actually maintain and fly aircraft operated by the airlines. It is not unusual for some maintenance functions, as well as other jobs related to airline operations, to be performed by a company other than the airline itself. Airlines have routinely contracted with other airlines for baggage handling, flight services, maintenance, and so forth. In addition, some of these functions are "outsourced" to specialized companies that are not themselves airlines. Food services and fueling, for example, are often done by nonairline companies. Some observers believe that maintenance outsourcing may be on the rise (Velocci, 1995:76; White, 1994:10), and others point to cyclical trends in maintenance outsourcing that are connected to periods of capacity expansion and contraction, affecting the airlines' ability to do maintenance work themselves (Fitzgerald, n.d.:17). To the extent that outsourced functions are performed by employees of another airline, then the employment data provided by the airlines are, in the aggregate, a good indication of job opportunities in those functions. To the extent that some of these functions are outsourced to nonairline companies, then these data may understate job opportunities. Furthermore, to the extent that such outsourcing is used to absorb variations in the need for services, then the employment data reported by the airlines may also understate the volatility in the demand for certain airline-related jobs.

New Hires

Hiring data on pilots have been most closely tracked over time by FAPA—formerly called Future Airline Pilots Association or Future Aviation Professionals of America—an Atlanta-based career and financial planning service for airline workers. FAPA data for the 10-year period ending in 1994 (Table 2-6) show that new pilot hires have ranged from a low of 3,256 in 1993 to a high of 13,401 in 1989. In 1993 the airlines classified by FAPA as globals/majors (equivalent to the carriers called majors by the FAA), the desired career destination for most professional pilots, hired only 491 people for their cockpit crews. During that

---

[7]The maintenance labor category is not precisely defined in the instructions (14 CFR Part 241) for completing Form 41, Schedule P-10. The instructions do indicate, however, that the category includes general maintenance workers as well as unallocated shop labor.

TABLE 2-6   Pilot Hires by Air Carrier Group and Year, 1985-1994

Carrier Type[a]	1985	1986	1987x	1988	1989x	1990x	1991	1992	1993	1994
Global/major	4,544		3,958	3,328	5,868	3,304	2,404	1,836	491	1,041
National	1,306		1,319	1,908	1,769	1,059	1,001	405	764	1,438
Turbojet	1,990		1,760	1,447	1,389	416	599	385	812	1,128
Turboprop	3,046	NA[b]	4,073	3,114	4,375	2,998	1,971	1,702	1,189	1,656
Total	10,886		11,110	9,797	13,401	7,777	5,975	4,328	3,256	5,263

[a]From 1985 to 1991 FAPA classified the four air carrier groups as major, national, jet, and regional. Data for these categories are included here under the headings global/major, national, turbojet, and turboprop, respectively. According to FAPA, these categories are generally comparable.

FAPA relies on the following air carrier designations and descriptions: global carriers have annual revenues of at least $5 billion, including at least $1 billion from international operations; major carriers have at least $1 billion in revenue; national carriers have annual revenues of less than $1 billion per year but at least $100 million; turbojet (scheduled) carriers have annual revenues of less than $100 million and fly primarily jet airplanes; turboprop (scheduled) passenger carriers have annual revenues of less than $100 million and fly primarily turboprop planes.

[b]FAPA sent only January-June data for 1986. In 1986, 5,226 pilots were hired in the first six months of the year.

SOURCE: Data provided by FAPA.

same year, FAPA reported that 1,615 pilots were on furlough from the major and other jet carriers (FAPA, 1994). These furloughed pilots would generally have the right to be recalled before the airline could hire new pilots; this recall right could last from five years to an unlimited period, depending on the stipulations of union contracts.

The new hire numbers in Table 2-6 can give a misleadingly high picture of pilot hiring unless one keeps in mind what FAPA calls "internal system defections," that is, pilots leaving one airline category for another (FAPA, 1995:2). As we discuss in Chapter 4 when we look at career pathways in aviation, pilots without military training often advance from smaller to larger planes within an airline and from smaller to larger airlines on their way to a cockpit job with a major carrier. We are unaware of any published data that indicate the true number of new pilots entering air carrier employment in any given year.

Annual hiring statistics for aircraft mechanics are less readily available than for pilots, but volatility appears to characterize this group as well. FAPA data show that U.S. airlines hired 12,893 mechanics in 1989 but only 1,407 in 1993 (Table 2-7).

TABLE 2-7   Maintenance Mechanic Hires by Air Carrier Group and Year, 1989-1995

Carrier Type[a]	1989	1990	1991	1992	1993	1994	1995
Global/major	9,018	5,358	1,130	812	341	386	560
National	1,508	1,028	650	414	524	539	852
Turbojet	883	523	278	195	190	390	198
Turboprop	1,484	1,412	858	759	352	477	411
Total	12,893	8,321	2,916	2,180	1,407	1,792	2,021

[a]From 1989 to 1991 FAPA classified the four air carrier groups as major, national, jet, and regional. Data for these categories are included here under the headings global/major, national, turbojet, and turboprop, respectively.

FAPA relies on the following air carrier designations and descriptions: global carriers have annual revenues of at least $5 billion, including at least $1 billion from international operations; major carriers have at least $1 billion in revenue; national carriers have annual revenues of less than $1 billion per year but at least $100 million; turbojet (scheduled) carriers have annual revenues of less than $100 million and fly primarily jet airplanes; turboprop (scheduled) passenger carriers have annual revenues of less than $100 million and fly primarily turboprop planes.

SOURCE: Data provided by FAPA.

Wages

A major reason people are concerned about making sure that the aviation industry has the employees it needs and that all individuals have equal access to jobs in this industry is that aviation jobs are widely perceived as good jobs—exciting, rewarding, and, perhaps most important of all, high paying. FAPA, for example, opens its guide to pilot employment by alerting the reader to the fact that flying "is also one of the few occupations, short of professional sports, that can make 'a mere employee' financially well-off. . . . [T]he major airline pilot enjoys a privileged position among American workers. It is a 'professional' role akin, in public image, to the roles of doctor and lawyer and compensated on a similar scale" (FAPA, 1993:1).

As mentioned earlier, part of the reason airlines have been perceived as paying high wages stems from the reduced incentives for cost control that existed under regulation. Part of the reason also stems from the high rate of productivity growth. Pilots in particular may have benefited from these factors. The outlook for airline salaries, in particular pilot salaries, may change in the future, however.

Deregulated competition has brought lower fares to passengers and has expanded travel, but it has also put pressure on the industry to change the terms and conditions of employment. In the post-deregulation era, airlines are not constrained to charge the same fares and are allowed to enter new markets, in most cases almost at will. There is now a strong incentive to reduce labor costs and to use lower labor costs as a competitive tool. If an airline can lower its labor costs, it can charge lower fares in its existing markets and enter into new markets.

Table 2-8 suggests that some adjustment in pilot salaries may already be taking place. In nominal terms, mean gross monthly earnings for captains increased about 28 percent between 1984 and 1995 in both the majors and the regionals. Adjusting for inflation (using the consumer price index), however, reveals that these earnings have actually fallen about 13 percent in real terms. First officer (copilot) salaries have fallen even faster in real terms among the majors but have increased among the regionals. This increase was from a very low base, however, as can be seen in the table.

Although it seems likely that the falling real earnings are in part an adjustment from earlier salary policies, it should also be noted that the industry was expanding during the period covered in the table. With expansion, more new pilots (and technicians) were being hired than were retiring, thereby bringing down the average seniority in the industry. Since earnings in the airline industry tend to rise with seniority, earnings growth could be expected to proceed more slowly during this period, even without an adjustment of wage policy.

The wage story is much the same among aircraft mechanics, as Table 2-9 shows. For mechanics in both the majors and the regionals, mean hourly earnings increased about 30 percent between 1984 and 1995. But when the adjustment is

TABLE 2-8 Mean Gross Monthly Earnings of Pilots (U.S. Certificated Carriers)

Carrier Type		Nominal		Inflation Adjusted		Consumer Price Index
		Captains	First Officers	Captains	First Officers	
Majors	1984	$8,953	$5,946	$8,953	$5,946	103.9
	1989	10,097	5,779	8,460	4,842	124.0
	1995	11,495	6,546	7,796	4,439	153.2
Regionals	1984	3,485	1,890	3,485	1,890	103.9
	1989	3,518	2,019	2,948	1,692	124.0
	1995	4,459	3,419	3,024	2,319	153.2

NOTE: Total monthly earnings include base pay and all other pay directly related to duty but exclude special allowances, such as those for room and board while away from the employee's home station.

SOURCES: Bureau of Labor Statistics (1985:Tables 1 and 2; 1990: Tables 1 and 2; 1995b: Table 1).

TABLE 2-9 Mean Straight-Time Hourly Earnings for Aircraft Mechanics

Carrier Type		Nominal	Inflation Adjusted
Majors	1984	$16.38	$16.38
	1989	16.96	14.21
	1995	21.32	14.46
Regionals	1984	11.22	11.22
	1989	13.07	10.95
	1995	14.82	10.05

NOTE: Earnings exclude premium pay for overtime and for work on weekends, holidays, and late shifts.

SOURCES: Bureau of Labor Statistics (1985:Table 9; 1990:Table 7; 1995b: Table 3).

made for inflation (again with the consumer price index), real wages are seen to have fallen about 12 percent for the majors and about 10 percent for the regionals.

Average earnings among employees in air transportation still significantly exceed those of workers in industry as a whole (Table 2-10). In 1992 average annual earnings for full-time employees of common carriers were $38,083, com-

TABLE 2-10 Average Annual Earnings per Full-Time Employee, Selected Years

Industry	1965	1970	1975	1980	1985	1990	1992
Transportation	$6,994	$9,391	$13,596	$20,677	$25,305	$28,916	$31,397
Railroad	7,462	10,112	15,363	25,358	36,746	41,814	49,706
Bus	5,550	6,875	9,299	13,224	15,813	19,676	20,539
Trucking/warehousing	6,625	8,672	12,709	18,864	22,291	25,833	27,828
Water	7,388	10,283	14,247	22,990	28,435	33,982	36,024
Air (common carrier)	8,496	12,027	17,084	25,498	31,789	34,890	38,083
Oil pipeline	8,053	10,765	16,765	26,182	36,947	43,474	50,947
Allied services	6,276	8,262	11,233	15,604	20,207	25,736	28,123
Manufacturing	6,566	8,381	11,903	17,996	24,549	29,746	32,370
Communications	6,820	8,752	13,726	21,388	31,381	38,382	42,076
Electric, gas, sanitary	7,476	10,023	14,056	21,917	31,669	38,930	42,998
Finance, insurance, real estate	5,971	7,821	10,619	15,864	23,724	31,682	36,159
All industry total	5,814	7,713	10,835	15,789	21,084	26,156	26,687

NOTE: These data are taken from the Department of Commerce's *Survey of Current Business*. Earnings represent total earnings, salaries, and wages, taken from the Wages and Salaries per Full-Time Equivalent Employee by Industry table.

SOURCE: Wilson (1994:60). Reprinted by permission.

TABLE 2-11   Average Annual Salary for Airline Pilots by
Category of Airline, U.S. Certificated Air Carriers, 1995

Carrier Type	Average Annual Salary		
	First Year	Sixth Year	Maximum
Global	$26,800	$87,600	$180,100
Major	29,000	69,400	150,200
National	22,800	41,000	79,100
Turbojet	29,800	46,800	66,700
Turboprop	14,000	29,600	45,100

SOURCE: Data provided by FAPA.

pared with $26,687 for all full-time industrial employees. The table also reveals that average salaries in air transportation before deregulation were at the top of the list of major industrial groupings, whereas they had fallen in relative terms more toward the middle of the pack by the early 1990s.[8] Indeed, rail, pipeline, communications, and electric, gas, and sanitary earnings all exceeded air transportation earnings in 1992, whereas all had been below them in earlier years.

Aviation's image as a high-salary field owes a great deal to the earnings of pilots; those employed by the majors have always been among the nation's leading wage earners. Here again, though, a closer look reveals a more complex story.

The most senior pilots flying the biggest jets continue to earn handsome salaries (Table 2-11). The maximum salary at the major airlines, calculated by averaging the maximum reported by each airline, was over $180,000 in 1995. A senior United pilot flying Boeing 747s earns over $200,000 (FAPA, 1995:3). A new recruit to a major airline, however, can expect a few low-earning years at the beginning, even with significant experience in other airlines or the military. Airline pilots at the large and medium regional airlines do noticeably less well than a senior captain at the majors. This wage difference takes on greater importance as regionals become less of a stepping-stone to the majors and more a source of permanent employment. Indeed, the rise in real wages for regional first officers may be partially due to this shift (Proctor, 1995:62; Transportation Research Board, 1996:49).

Individual pilot salaries, along with other aspects of employment, such as

---

[8] A Department of Transportation study that also reports on average wages in air transportation and selected other industries notes that the Bureau of Labor Statistics, the source of the data, does not publish air transportation averages separately because the small size of the sample result in statistics that do not meet its standards for reliability. The authors note, however, that the long-term time trend of these data do seem reliable (U.S. Department of Transportation, 1992:29-30).

advancement from first officer to captain, assignments to the most desirable aircraft, and bidding for vacation and monthly trip schedules, are governed strictly by the pilot's seniority number, which measures time in service with the current employer. Pilots lose their seniority numbers if they shift from one employer to another, which significantly affects their earning prospects.

Earnings of pilots who don't work for airlines vary considerably. A 1995 survey conducted by *Professional Pilot* magazine found that "corporate pilot salaries wander way above and below the median but there is no logic to the pay rates" and further noted that "there are many factors other than straight aircraft type that determine salaries."[9] The survey reported average salaries for corporate jet pilots flying different types of jet aircraft as ranging between $60,000 and $103,000 for chief pilots, between $50,000 and $90,000 for captains, and between $36,000 and $69,000 for first officers. The salaries for corporate turboprop pilots in different types of aircraft ranged between $41,000 and $51,000 for chief pilots, between $37,500 and $50,000 for captains, and between $25,000 and $32,600 for first officers (*Professional Pilot*, 1995:68-69).

For other types of commercial flying jobs, salary levels are lower. In the early 1990s the FAA reported average midrange salaries for commercial pilots (e.g., patrol, ferry, helicopter, aerial survey) as $20,000 annually, for agricultural pilots as $17,000 annually, and for air taxi/charter pilots as $14 per hour (Federal Aviation Administration, n.d.(a):2).

Licensing and Certification

Pilots and aviation mechanics, as well as certain other aviation workers such as air traffic controllers, are subject to a series of licensing requirements established by the federal government.

Pilots

Pilots must be certified before they can operate an aircraft. Part 61 of the Federal Aviation Regulations (14 CFR 61) prescribes the requirements for certifying pilots and flight instructors and recognizes five types of pilot certification: student pilot, recreational pilot, private pilot, commercial pilot, and airline transport pilot. Part 61 also provides for a series of ratings, including aircraft type ratings that permit pilots to operate specific kinds of equipment, that can be applied to the various types of certificates (Table 2-12). It further establishes the conditions under which the various certifications and ratings are necessary.

---

[9]According to a note accompanying the 1995 salary survey results, the reported data are based on reader responses combined with data from within the industry and should be used as a guide only, since salaries are affected more by individual company, size of the flight department, and years of company service than by aircraft size.

TABLE 2-12   Pilot Certificates and Selected Ratings

Certificates
    Pilot certificates
        Student pilot
        Recreational pilot
        Private pilot
        Commercial pilot
        Airline transport pilot
    Flight instructor certificate

Ratings: placed on pilot certificates (except student pilot) where applicable
    Aircraft category ratings
        Airplane
        Rotorcraft
        Glider
        Lighter-than-air
    Airplane class ratings
        Single-engine land
        Multiengine land
        Single-engine sea
        Multiengine sea
    Aircraft type ratings: listed in an FAA advisory circular, they include
        Large aircraft, other than lighter-than-air
        Small turbojet-powered airplanes
        Small helicopters for operations requiring an ATP certificate
        Other aircraft type ratings specified by the FAA administrator
    Instrument ratings (on private and commercial pilot certificates only)
        Instrument—airplanes
        Instrument—helicopter

SOURCE: 14 CFR 61.5.

The rules governing the airlines (Federal Air Regulations Parts 121 and 135) determine the minimum licenses and ratings that their pilots must hold. For example, pilots flying in command of a Part 121 aircraft must have an airline transport pilot (ATP) certificate with the appropriate aircraft type rating. The minimum requirement for a Part 121 aircraft pilot (not flying in command) is a commercial pilot certificate with an instrument rating. Many Part 135 pilots in command, including all commuter pilots in command, are also required to hold ATP certificates and appropriate aircraft type ratings.

    Qualifying for an ATP certificate requires a pilot to meet certain basic eligibility standards: to be at least 23 years of age; have "good moral character"; be able to read, write, and understand English and speak it without accent or impediment that would interfere with two-way radio communication; be a high school graduate; and hold a current first-class medical certificate, the requirements for which are outlined in another part of the regulations (Part 67). The certificate

TABLE 2-13   Certificates and Ratings Held by Newly Hired Air Carrier
Pilots, 1994

Certificate/Rating	Globals/Majors	Nationals	Turbojets	Turboprops
Air transport pilot (ATP)	97%	76%	91%	53%
FE or FEw	88	37	60	40
Commercial and instrument only	0	17	5	28
CFI/CFII	23	50	31	70
Type rating	60	28	51	12

NOTE: FE = flight engineer certificate; FEw = flight engineer written exam only; CFI = certified flight instructor; CFII = certified flight instructor with instrument rating.

SOURCE: FAPA (1995:10).

also requires that the pilot pass written and flight tests and have specified levels of aeronautical experience. These include at least 250 hours of flight time as pilot in command of an airplane, or as copilot performing the duties and functions of a pilot in command under the supervision of a pilot in command, or some combination thereof. He or she must also have at least 1,500 hours of flight time as a pilot. Commercial pilot certificates have lower requirements, notably a current second-class medical certificate and 250 hours of flight time as a pilot. As we discuss in Chapter 4, an expensive aspect of becoming an airline pilot can be the costs of accumulating the flight time necessary to acquire the required certificates.

All pilots must hold certificates and ratings as prescribed in federal regulations to fly planes, but how many of the qualifications they need in order to be hired by an airline is very much dependent on supply and demand in the pilot labor market. Currently, major airlines are able to demand that most new hires already hold the ATP certificate, and the flight experience and ratings they require vastly exceed the minimums required for that certificate (Table 2-13). Although some newly hired airline pilots typically have had type ratings when they were hired, it has been more common to obtain them through company-sponsored training once on the payroll. Southwest, however, requires a type rating in order for pilots to be hired. By contrast with current airline hiring practices, there have been times of tight supply in the past when new-hire qualifications were significantly lower, including a period in the 1960s when United even hired so-called "no-time" pilots and provided extensive training for them. We return to the issue of pilot certifications and ratings, who provides them and what they cost, in Chapter 4.

Technicians

Certification for "airmen other than flight crewmembers" is specified in

Federal Air Regulations Part 65. Occupations covered include air traffic control tower operators, aircraft dispatchers, mechanics, repairmen, and parachute riggers. We focus here on mechanics and repairmen; mechanics are increasingly being described as aviation maintenance technicians (AMTs), and we use the terms interchangeably in this report.

AMTs are responsible for keeping aircraft in an airworthy condition on a day-to-day basis. Airframe specialists work on all aircraft parts except instruments, power plants, and propellers. Power plant technicians work on engines and do some limited work on propellers. Avionics experts maintain aircraft navigation and communication radios, weather radar systems, autopilots, and navigation, engine, and other instruments and computers.

Part 65 provides for mechanics certification with two possible ratings (airframe and power plant) and for repairman certification. Most AMTs have the mechanics certification with both ratings, which is commonly referred to as an airframe and power plant (A&P) certificate. The A&P exam includes three written tests and three oral/practical tests. Individuals who work on avionics may have an A&P certificate along with appropriate licenses from the Federal Communications Commission. Repairmen do not have individual certification they can take with them from employer to employer. An FAA-certified repair shop employs specialists in such areas as welding, metal forming, and engine repair; a specialist can obtain a repair certificate after 18 months of practical experience in a specific job.

Unlike pilots, maintenance technicians do not need to be certified to work on aircraft, although they are prohibited from undertaking certain responsibilities, such as approving the return of an aircraft to the flight line, unless they are certified. FAPA cites an FAA survey indicating that approximately 75 percent of the new AMT hires at the major airlines have the A&P certificate, as do 98 percent of new hires at the regional airlines who fly turbojet aircraft (White, 1994:13).

Federal certification for aviation maintenance technicians and repairmen appears to be on the verge of major changes, designed to recognize the increasing skills necessary to work on today's aircraft. After completing the first full regulatory review of the certification requirements for aviation maintenance personnel since 1962, the FAA has proposed a new Part 66 of the Federal Air Regulations, which will focus only on maintenance personnel and will replace the maintenance provisions in subparts D and E of Part 65. These proposed regulations call for the creation of additional certificates and ratings and expand current certification requirements (Federal Aviation Administration, n.d.(c):1).

Under the new Part 66,[10] the term *mechanic* will be replaced by *aviation*

---

[10]As the committee finishes its work, the proposed Part 66 is still being reviewed by the FAA. There may be changes in the provisions described here before the new rules receive final clearance.

maintenance technician; *repairmen* will become *aviation repair specialists*. Two new certificates will be created in place of the existing mechanics certificates: aviation maintenance technician (AMT) and aviation maintenance technician (transport) or AMT(T). After the new regulations are implemented, individuals with the knowledge, skills, and experience traditionally displayed by holders of A&P certificates will be issued AMT certificates. AMTs will not, however, be able to approve transport category aircraft for return to service. This responsibility will be limited to holders of AMT(T) certificates, which will require additional preparation and training. These changes are designed to respond to the findings of the Pilot and Aviation Maintenance Technician Blue Ribbon Panel, which "concluded that existing certification requirements did not give aviation maintenance personnel the entry-level experience and skills necessary to perform work involving transport category aircraft that use new technology" (Federal Aviation Administration, n.d.(c):4). AMTs and ATMT(T)s will be required to register periodically with the FAA, providing for the first time a means for determining the number and location of aviation maintenance personnel.

A new "portable" aviation repair specialist (ARS-I) certification will be available to repair personnel. This certification will be issued to individuals, unlike the current repairman certificates held by aviation maintenance organizations. The ARS-I certificate will be based on uniform national standards, again, unlike the repairman certificates that they replace.

Diversity and the Aviation Workforce

Aviation occupations, although changing, do not mirror the diversity of the overall American workforce. Although aviation employees as a group are not dramatically different in sex, race, and ethnic makeup from all employees, the representation of women and racial minorities varies substantially from occupation to occupation. Pilots and senior managers continue to be predominantly white and male; mechanics are less likely to be white than are pilots and managers but are mostly men. As we explain in more detail in Chapters 3 and 5, these employment patterns are in part the result of a history of explicit and implicit policies against hiring women and minorities for aviation jobs in the military and at the airlines, policies that have been the subject of legal challenge and government investigation for several decades. The aviation workforce is still affected by a history of discrimination. Although substantial progress has been made, concerns about discrimination still exist.

Although our generalizations about the sex and racial/ethnic composition of the aviation workforce are not controversial, they are somewhat complicated to illustrate statistically because of limitations in available databases. Data from the Equal Employment Opportunity Commission (EEOC) (Table 2-14) give an overview of employment in broad occupational areas by industry and by racial/ethnic group and sex in 1993, based on a partial census of private-sector employment.[11] These data indicate that the scheduled air transport industry does somewhat better at employing women than transportation and public utilities overall, although it still falls below the proportion of women employed in all private industry. Scheduled air transport does somewhat less well in employing members of minority groups than the broader transportation and public utility industry. The range of variation across broad occupational groupings is wide. Because the EEOC occupational categories are defined quite generally and are uniform across all industries, it is not possible to isolate information from these data on the narrower aviation occupations of particular interest in this study. We can only point out that pilots are included in the "professional" category and mechanics in "craft workers."

More specific demographic information on pilots and mechanics (but not on aviation management) can be obtained from Census Bureau surveys. Here, a different data problem emerges: given the sample sizes involved, the numbers of pilots and mechanics are so small that monthly and annual estimates from the Current Population Survey suffer from standard errors large enough to render them largely useless for our purposes. For the same reason, we do not have confidence that we can obtain a reliable estimate of changes over time from this source, despite the fact that the Current Population Survey does publish monthly and annual estimates of the percentages of pilots and mechanics who are women and minorities.

Therefore, in terms of national data collected by agencies adhering to high statistical standards, we are dependent primarily on the decennial census of the U.S. population for information about the demographic characteristics of those who report their occupation as pilot or mechanic.[12] In 1980, 96 percent of pilots and 86 percent of aircraft mechanics were white men (Table 2-15). The comparable percentage for the civilian labor force 16 years of age and older was 50 percent. By 1990, the pilot and mechanic workforces were marginally more diverse (Table 2-16): 92 percent of pilots and 76 percent of aircraft mechanics were white men, compared with 43 percent of the civilian labor force. Less than

---

[11]EEOC data cover employers of 100 or more employees and federal government contractors with 50 or more employees and contracts of $50,000 or more. The data reported in Table 2-14 are based on EEOC establishment reports and for multiestablishment firms exclude establishments with fewer than 50 employees.

[12]The FAA reports on the number of pilot and mechanic certificates held by women but not by racial minorities. The number of active pilot certificates held probably bears some relationship to the number of individuals employed as pilots or available for employment, since keeping these certificates current requires a valid medical certificate that (for the air transport pilot certificate) must have been issued within the last 6 months. We look at these statistics in Chapter 5. The number of "active" mechanics certificates, however, represents all certificates ever issued and therefore reveals little about the current workforce or about qualified individuals available for employment.

TABLE 2-14   Occupational Employment in Private Industry by Racial/Ethnic Group and Sex, 1993

	Total Employment	Officials and Managers	Professionals	Technicians
All employees	100.0%	100.0%	100.0%	100.0%
Female	46.5	29.9	50.2	46.9
Minority	23.5	10.8	14.3	19.9
Black	12.7	5.3	5.5	10.5
Hispanic	7.2	3.1	2.7	4.7
Transportation				
& public utilities	100.0	100.0	100.0	100.0
Female	32.8	26.0	26.5	19.4
Minority	22.0	13.4	13.1	16.5
Black	13.2	7.5	5.6	8.8
Hispanic	6.2	3.6	3.1	4.8
Scheduled air				
transportation	100.0	100.0	100.0	100.0
Female	42.0	27.2	8.8	28.6
Minority	19.5	13.1	5.7	12.0
Black	9.5	6.6	1.9	6.1
Hispanic	5.7	3.2	1.9	2.5

NOTE: A total of 370,000 employees in scheduled air transport are covered by this survey.

SOURCE: Equal Employment Opportunity Commission (1994:Table1).

2 percent of pilots were black, less than 3 percent were of Hispanic origin, and in both cases these were mostly men. Overall the proportion of women reporting their occupation as pilot grew from 1.4 to 3.6 percent over the decade. The proportion of female mechanics grew, but more slowly, from 3.3 to 4.7 percent. Black and Hispanic men were noticeably more likely to report their occupation as aircraft mechanic in 1990 than their representation in the civilian labor force would suggest, but black and Hispanic women, like white women, were not very likely to be aircraft mechanics.

It is very difficult to find statistical data on two related issues of interest to the committee: the representation of women and minorities in the workforces of the major airlines and the proportion of new hires that are women and minorities. A recent article noted that "most airlines are unwilling to provide data on the gender or ethnicity of their work forces" (Henderson, 1995:34). Some evidence comes from private aviation interest groups. The International Society of Women

Sales Workers	Office and Clerical Workers	Craftworkers	Operatives	Laborers	Service Workers
100.0%	100.0%	100.0%	100.0%	100.0%	100.0%
56.5	82.8	11.4	32.2	34.4	55.3
20.0	23.7	17.9	29.7	38.5	40.2
10.6	14.0	9.2	17.1	18.7	23.8
6.8	6.4	6.5	9.2	16.3	12.4
100.0	100.0	100.0	100.0	100.0	100.0
62.4	77.3	5.7	11.7	15.8	60.7
25.1	27.9	16.1	25.8	37.0	27.5
14.5	18.4	9.0	16.6	22.0	16.1
7.5	6.9	5.4	7.4	11.7	7.7
100.0	100.0	100.0	100.0	100.0	100.0
72.8	67.1	6.2	10.9	34.9	76.8
24.7	21.5	17.3	27.8	48.3	21.4
13.2	11.6	7.3	14.2	23.6	10.4
6.3	5.8	5.7	8.8	13.6	6.4

Airline Pilots estimates that the pilot workforce at the average airline is about 5 percent female.[13] FAPA reports that 6.8 percent of the pilots hired by U.S. major and global airlines in 1994 were female. The Organization of Black Airline Pilots, a professional association, found in a 1994 survey (Table 2-17) that 1.2 percent of all the pilots at the largest U.S. airlines were black and 2.3 percent were white females (Henderson, 1995:34-37). Since pilots without military training would normally work for national and regional airlines before being hired by the majors, one would expect that these airlines might have larger percentages of female and minority pilots, hired in more recent years as employment barriers have diminished. Unfortunately, we found no data that would verify this expectation. Similarly, we know of no data about the numbers of women and minorities at various levels of management in the airline industry, although the numbers

---

[13]This estimate may be somewhat high, given that women hold fewer than 5 percent of commercial and air transport pilot certificates (see Table 5-1).

TABLE 2-15   Detailed Occupation of the Experienced Civilian Labor Force by Sex, Race, and Spanish Origin, 1980 (percentages in parentheses)

	White		Black	
	Male	Female	Male	Female
Experienced civilian labor force 16 years and over	51,781,293 (49.8)	37,178,485 (35.7)	5,276,500 (5.1)	5,180,564 (5.0)
Airplane pilots and navigators	73,572 (96.6)	986 (1.3)	652 (0.9)	26 (0.0)
Aircraft mechanics	96,780 (85.7)	2,892 (2.6)	6,703 (5.9)	638 (0.6)

NOTE: Experienced civilian labor force consists of the employed and the experienced unemployed.

[a]Persons of Spanish origin may be of any race.

SOURCE: Bureau of the Census (1984:Table 277).

TABLE 2-16   Detailed Occupation of the Civilian Labor Force by Sex, Race, and Hispanic Origin, 1990 (percentages in parentheses)

			Not of Hispanic origin			
	Hispanic origin[a]		White		Black	
	Male	Female	Male	Female	Male	Female
Civilian labor force 16 years and over	5,800,180 (4.7)	4,133,543 (3.3)	52,652,638 (42.6)	43,590,483 (35.3)	6,108,277 (4.9)	6,727,324 (5.4)
Airplane pilots and navigators	2,273 (2.1)	88 (0.1)	100,624 (91.6)	3,450 (3.1)	1,594 (1.5)	292 (0.3)
Aircraft mechanics	14,031 (8.4)	623 (0.4)	126,013 (75.7)	5,604 (3.4)	12,176 (7.3)	1,220 (0.7)

[a]Persons of Hispanic origin may be of any race.

SOURCE: Bureau of the Census (1992:Table 1).

Am. Indian, Eskimo, Aleut.		Asian and Pacific Islander		Spanish origin[a]		Total	
Male	Female	Male	Female	Male	Female	Male	Female
331,865	248,765	943,556	820,369	3,572,520	2,373,701	59,753,512	44,304,473
(0.3)	(0.2)	(0.9)	(0.8)	(3.4)	(2.3)	(57.4)	(42.6)
285	—	451	13	1,194	78	75,149	1,039
(0.4)	(0.0)	(0.6)	(0.0)	(1.6)	(0.1)	(98.6)	(1.4)
786	72	2,441	51	7,636	280	109,124	3,740
(0.7)	(0.1)	(2.2)	(0.0)	(6.8)	(0.2)	(96.7)	(3.3)

Am. Indian, Eskimo, Aleut.		Asian and Pacific Islander		Other race		Total	
Male	Female	Male	Female	Male	Female	Male	Female
426,376	365,896	1,864,689	1,631,072	46,041	38,931	66,986,201	56,487,249
(0.3)	(0.3)	(1.5)	(1.3)	(0.0)	(0.0)	(54.3)	(45.7)
384	19	1,021	48	33	—	105,929	3,897
(0.3)	(0.0)	(0.9)	(0.0)	(0.0)	(0.0)	(96.5)	(3.5)
1,102	82	5,275	264	96	—	158,693	7,793
(0.7)	(0.0)	(3.2)	(0.2)	(0.1)	(0.0)	(95.3)	(4.7)

TABLE 2-17   Black and Female Airline Pilots, 1994

Airline	Black		Female		Total
American	70	0.7%	175	1.8%	9,928
AMR Eagle	20	0.8	6	0.2	2,400
Continental	10	0.2	40	1.0	4,100˙
Delta	54	0.6	35	0.4	9,406
FedEx	64	2.9	99	4.4	2,234
Northwest	28	0.5	75	1.4	5,239
Southwest	24	1.1	47	2.7	1,715
TWA	26	0.8	44	1.4	3,147
United	212	2.7	500	6.2	8,015
UPS	55	3.7	82	5.5	1,503
USAir	59	1.0	115	2.0	5,709
Total	622	1.2	1,218	2.3	53,396

SOURCE: Unpublished data from the Organization of Black Airline Pilots, cited in Henderson (1995:37).

near the top are generally conceded to be few (Henderson, 1995:33; Petzinger, 1995).[14]

WORKER SUPPLY AND DEMAND IN THE AVIATION INDUSTRY

Aviation is an industry characterized by wide gyrations in hiring in its most specialized occupations: pilots and AMTs. The supply of personnel, too, can be affected by short-term discontinuities, such as war, and industrial restructuring that results in major layoffs or airline closures. Following World War II and the Korean War, many trained pilots and mechanics were unable to find jobs in aviation. During the mid-1960s and the late 1980s, hiring levels grew rapidly, accompanied by concerns about the industry's ability to find the skilled workers it needed. During the growth period of the 1960s, the FAA established a board to review the industry's manpower requirements; similarly, in the late 1980s, Congress initiated a review that resulted in the Pilot and Aviation Maintenance Tech-

---

[14]We are also uncertain about the reliability of the estimates provided by private aviation groups. They generally do not report on their survey methodology or response rates. FAPA indicates that it obtains overall pilot hiring numbers directly from the air carriers; however, other FAPA data, which we and others use extensively to learn about pilot hiring (such as sex, military background, education), appear to come from surveys distributed to successful job applicants by FAPA members hired at the same airline. The proportion of newly hired pilots completing FAPA surveys varies greatly from year to year and among carrier groups.

nician Shortage Blue Ribbon Panel's 1993 report. Even as the Blue Ribbon Panel conducted its work in the early 1990s, however, the aviation industry's fortunes reversed, and its labor market went from perceived shortage to an oversupply of trained and available pilots and mechanics (Blue Ribbon Panel, 1993).

The issue of supply and demand is further complicated by the question of what constitutes a "qualified" candidate for a pilot or AMT position. To fly or work on an aircraft, an individual must meet the minimum conditions laid down by the FAA. These minima are generally far below the standards that the major airlines expect successful job applicants to meet and, for the most part, have been able to demand even in periods when labor markets were tight. When, during the peak hiring periods of the 1980s, the major airlines reduced "soft" requirements for pilots, such as age, education, and vision requirements, they were able to attract the new hires they needed without much variation in their standards for total flight time, certificates, and experience levels (Blue Ribbon Panel, 1993:15). In fact, experience levels tended to be high relative to the FAA-mandated minima, which reduced the amount of training the airlines themselves needed to provide.

Other factors also make it difficult to determine when the supply of labor is adequate. Training does not stop at a fixed point, but rather extends across a continuum of training needs. Pilots in particular undergo formal training throughout their careers, from early post-hire training to familiarize them with a particular airline's operating procedures to recurrent and upgrade training to maintain their skills and allow them to move from one type of aircraft to another. Whereas airlines normally pay for recurrent and upgrade training for their current workforce, they face a trade-off at the entry level between requiring training or experience as a condition of hire (thereby putting more of the responsibility on the individual applicant) and providing training after hire (when the airline will have to bear the cost). These factors make statistical forecasts about the new-hire labor market highly dependent on the forecaster's judgment about the nature of a "qualified" applicant pool. The difficulties are compounded by problems of forecasting supply and demand for an industry whose fortunes are as closely tied to broader economic conditions and cycles as the airline industry. Technological changes further complicate the picture; increasingly sophisticated aircraft are becoming easier to fly (perhaps reducing the qualifications that pilots will need) but more complicated to maintain and repair (suggesting that requirements for AMT certification will increase). These complexities contribute to sometimes ambiguous forecasts, such as those of the Blue Ribbon Panel, which found that beyond the near term there may be a shortage of entry-level pilots "who meet the *qualification and experience* standards currently accepted" and of "fully qualified AMTs" able to meet "the rapidly increasing demands of technology" (Blue Ribbon Panel, 1993:51 and 60-1).

From the committee's perspective, our interest is less in making numerical predictions than in assessing the capacity of aviation and its training affiliates to adjust to whatever future labor market conditions may develop, and in considering how cyclical swings in demand may affect efforts to bring more women and minorities into the aviation workforce.

RECAPITULATION

Some important themes emerge from this overview. First, for most of the history of civilian aviation, the federal government was a dominant force. Although dramatic changes have occurred since deregulation, the industry and its workforce still retain much of the basic shape molded by government controls. For example, were it not for the prior role of the CAB, the salaries of pilots would probably not be as high. Licensing and certification requirements—a major determinant of training demands—are still prescribed by the federal government.

Second, the aviation industry—a major contributor to the economy in its own right—is greatly affected by broader economic trends, which in turn affect its workforce policies. For example, the specific combinations of training, credentials, prior experience, and other qualifications required for pilot, AMT, and other key jobs may vary according to the airlines' perceptions of supply and demand.

Third, factors that made piloting, in particular, something of a dream job have changed. Pilots are only a small fraction, although a highly visible fraction, of the aviation workforce. There are many other jobs in aviation that may also be appealing to people interested in flight.

Fourth, the committee faced serious problems in its quest for reliable data to answer key questions in its charge, such as the numbers of women and minorities in various aviation occupations. This theme emerges again in Chapter 4, as we encountered similar difficulties in painting a statistical portrait of civilian aviation training.

3

The Impact of Military Downsizing

Historically, the military has been an important source of trained professionals for commercial aviation, especially pilots. Any attempt to assess training issues in the industry, therefore, requires an understanding of how the military influences the supply of aviation specialists.

Since at least the late 1980s, the military services have been undergoing a fundamental reshaping, restructuring, and drawing down of force size. These efforts have reduced the inventory of officers and enlisted personnel in aviation-related occupations in the services and have lowered the numbers of individuals being recruited into these specialties. Such reductions will affect the ability of the civilian air carriers to draw on the military for trained aviation personnel. The major air carriers, faced with a dwindling supply of trained military pilots and mechanics, will have to meet their future hiring needs by relying to a greater extent than they presently do on civilian sources of supply. (Smaller air carriers already draw many pilots from civilian ranks.)

As a major training ground for pilots and aviation maintenance technicians, the military's aviation-related workforce is not noticeably more diverse than the civilian aviation workforce. Minorities and women are better represented in military aviation specialties than they used to be, but (with the exception of minority male mechanics) their presence in these jobs is small, still significantly lagging their representation in the overall population. Because the proportion of minority and women pilots in particular is very low, the military drawdown will not have much effect on the diversity of the pool of trained pilots available to the air carriers. The drawdown does, however, mean that opportunities are shrinking for minorities and women, as well as for white men, to receive aviation training by joining the armed forces.

THE MILITARY AS A SUPPLIER OF PILOTS AND TECHNICIANS

Several studies have highlighted the importance of the military as a source of pilots for the commercial air carriers (Blue Ribbon Panel, 1993:31-33; Levy, 1995:25, 29; Thie et al., 1994:34-37, 1995:9). Historically, the major carriers have relied on the military for about 75 percent of their pilots (Levy, 1995:22). A look beyond the major carriers and a focus on recent years, however, reveals a more complicated picture.

Table 3-1 shows the number of pilots hired by U.S. airlines from 1985 through 1994 (except for two years for which data are missing) and the percentage of these new hires who had military backgrounds. As the table indicates, for the largest airlines (those designated global/major by FAPA), the share of new hires with military backgrounds has been above 75 percent in the 1990s. In 1987 and 1988, however, it was below 70 percent, and in 1985 it was below 50 percent. Overall, for the years represented in the table, an average of 68 percent of new hires had military backgrounds. For the national carriers, however, the overall share with military backgrounds was only 40 percent, and for the turbojets and turboprops, the figures are even lower. The turboprops averaged fewer than 20 percent of pilots with military experience during the period represented in the table. Thus the dominance of pilots with military background has been a feature only of the largest airlines, and even for them, it has varied from year to year.

An examination of the variation in year-to-year hiring statistics reveals an interesting pattern. Figure 3-1 shows, for the global/major airlines, the percentage of pilots with military backgrounds plotted against the number of pilots hired in that year. Figure 3-2 shows the same plot for the national airlines. In both cases, there is a clear inverse pattern. When the number of pilots hired is low, the percentage with military backgrounds is high. Conversely, when the number of pilots hired is high, the percentage with military backgrounds is lower. This pattern holds for both the global/major airlines and the national airlines. A comparison of the two figures reveals that the nationals generally hire fewer pilots, but that, even when the two groups are hiring about the same number of pilots, the percentage with military backgrounds is higher for the global/majors than for the nationals.

Although these graphs may not be conclusive, when combined with anecdotal evidence, they are suggestive of a couple of airline hiring patterns. The first pattern is that airlines seem to prefer pilots with military backgrounds. When hiring relatively few pilots, airlines will hire from military ranks. When airline hiring needs are larger, however, they seem quite willing to reach into the civilian ranks to fill their pilot needs. The second apparent pattern is that pilots would generally prefer working for larger airlines than for smaller airlines. As result,

TABLE 3-1 Pilot New Hires, Selected Years

Carrier Type[a]	1985	1987	1988	1990	1991	1992	1993	1994
Number hired								
Global/major	4,544	3,958	3,328	3,304	2,404	1,836	491	1,041
National	1,306	1,319	1,908	1,059	1,001	405	764	1,438
Turbojet	1,990	1,760	1,447	416	599	385	812	1,128
Turboprop	3,046	4,073	3,114	2,998	1,971	1,702	1,189	1,656
Total	10,886	11,110	9,797	7,777	5,975	4,328	3,256	5,263
Percentage with military background								
Global/major	47	66	69	76	80	85	93	82
National	33	40	38	38	46	58	48	34
Turbojet	24	39	41	39	45	51	25	64
Turboprop	16	17	18	22	14	25	13	22

[a]Data for 1986 and 1989 are incomplete and are not included. For the period 1985 to 1991, the following air carrier categories were used by FAPA: major, national, jet, and regional. For purposes of this table, data for those years are included, respectively, in the following categories: global/major, national, turbojet, and turboprop. Pilots with military backgrounds may also have civilian experience.

FAPA relies on the following air carrier designations and descriptions: global carriers have annual revenues of at least $5 billion, including at least $1 billion from international operations; major carriers have at least $1 billion in revenue; national carriers have annual revenues of less than $1 billion per year but at least $100 million; turbojet (scheduled) carriers have annual revenues of less than $100 million and fly primarily jet airplanes; turboprop (scheduled) passenger carriers have annual revenues of less than $100 million and fly primarily turboprop planes.

SOURCE: Data provided by FAPA.

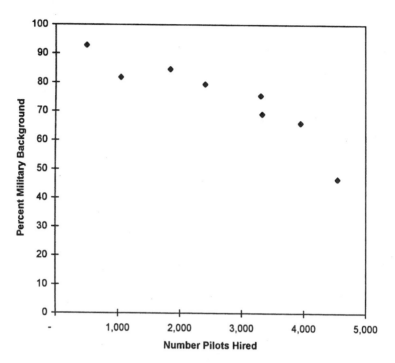

FIGURE 3-1 Number of pilots hired by global/major airlines and percentage with military background. SOURCE: Data provided by FAPA .

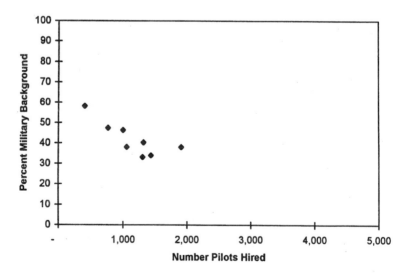

FIGURE 3-2 Number of pilots hired by national airlines and percentage with military background. SOURCE: Data provided by FAPA.

larger airlines are able to hire a higher percentage of military pilots than are smaller airlines.

There is less information about the military as a source of civilian aircraft maintenance personnel than there is about pilots, and the information that exists is somewhat contradictory. The FAA's Pilot and Aviation Maintenance Technician Blue Ribbon Panel (Blue Ribbon Panel, 1993:19) found that the military supplied a much lower percentage of technicians than pilots to the air carriers— an average of 23 percent of major carriers' new hires, based on results from a 1992 survey. FAPA contends that "[t]he military services traditionally supply a strong percentage of maintenance technicians to civilian aviation" and that the military services "provide the *majority* of skilled avionics technicians to commercial aviation" (FAPA, 1994:11, emphasis in original). To the extent that civilian training paths already provide a higher proportion of aviation maintenance technicians than they do pilots, then civilian aviation is less likely to be affected by reductions in the supply of technicians receiving military training.

The military drawdown is not reducing all parts of the military equally. This can be seen by looking in more detail at the aviation-related occupations in the armed services.

The principal military aviation occupations for pilots are found in the officer corps and are included under tactical operations (pilots and crews and operations staff officers). For aviation maintenance technicians, some of the military occupations are found in the officer corps under engineering and maintenance (design, development, production, and maintenance engineering officers), and some are found among enlisted personnel who work in electrical and mechanical repair (maintenance and repair of electrical, mechanical, hydraulic, and pneumatic equipment) (U.S. Department of Defense, 1994).

For officers, the specific occupational categories of interest are:

- fixed-wing fighter and bomber pilots (pilots of various types of fighter, attack, and bomber aircraft);
- other fixed-wing pilots (non-fighter and bomber fixed-wing pilots such as those engaged in transport, supply, and reconnaissance);
- helicopter pilots;
- aircraft crews (navigators and other officer aircraft crew personnel); and
- aviation maintenance officers.

For enlisted personnel, the specific category of interest is aircraft repair and maintenance (aircraft engines, electrical systems, structural components and surfaces, and launch equipment).

Table 3-2 summarizes the number of people in these major categories in 1980, 1985, 1990, and 1994. The military drawdown can be seen in the table. The number of officers in 1994 was 17 percent lower than in 1990 and 20 percent

TABLE 3-2 Active-Duty Personnel in Aviation-Related Occupations, All Services, Selected Years

Occupation	1980	1985	1990	1994	1990-1994 Percent Change	1985-1994 Percent Change	1980 Share	1994 Share
All officers	278,014	309,137	296,886	245,841	-17	-20	100	100
Tactical operations	86,722	116,037	121,110	95,707	-21	-18	31.2	38.9
Fixed-wing pilots	25,597	26,862	24,591	21,476	-13	-20	9.2	8.7
Army	597	549	436	413	-5	-25	0.2	0.2
Navy	7,394	6,945	6,563	4,966	-24	-28	2.7	2.0
Air Force	16,132	17,585	16,147	14,634	-9	-17	5.8	6.0
Marines	1,474	1,783	1,445	1,463	1	-18	0.5	0.6
Helicopter pilots	12,629	20,276	18,690	16,410	-12	-19	4.5	6.7
Aircraft crews	11,966	12,525	12,181	10,758	-12	-14	4.3	4.4
Engineering and maintenance	44,574	48,325	39,548	32,383	-18	-33	16.0	13.2
Aviation maintenance	8,028	8,069	7,388	6,412	-13	-21	2.9	2.6
Other occupations	146,718	144,775	136,228	117,751	-14	-19	52.8	47.9
Enlisted personnel	1,758,658	1,828,282	1,732,414	1,350,681	-22	-26	100	100
Electrical/mechanical repair	348,461	368,245	347,765	269,820	-22	-27	19.8	20.0
Aircraft and aircraft related	146,290	151,640	135,610	108,699	-20	-28	8.3	8.0
Other occupations	1,410,197	1,460,037	1,384,649	1,080,861	-22	-26	80.2	80.0

NOTE: Data shown are for the end of the defense fiscal year (September). Data for occupational categories correspond to the following occupational codes: tactical operations, 2; fixed-wing pilots, 2A, 2B; helicopter pilots, 2C; aircraft crews, 2D; engineering and maintenance, 4; aviation maintenance (enlisted only), 600, 601, 602, and 603 (U.S. Department of Defense, 1994).

SOURCE: Unpublished data provided by Defense Manpower Data Center, U.S. Department of Defense.

lower than in 1985. Similarly, the number of enlisted personnel was 22 percent lower than in 1990 and 26 percent lower than in 1985.

Fixed-wing pilots were down 13 percent between 1990 and 1994 and 20 percent between 1985 and 1994. Fixed-wing pilots made up a smaller share of the officer corps (8.7 percent) in 1994 than they did in 1980 (9.2 percent). Helicopter pilots are listed in the table, but there is no strong evidence that they are a major source of pilots for the airlines.

Changes in the numbers of pilots are not new phenomena in the military. Pilot inventory and requirements have fluctuated significantly over the past 40 years, both up and down (Thie et al., 1995:3). For the period of 1950 to 1990, the Air Force pilot inventory escalated from about 24,000 in 1950 to almost 55,000 in 1957, then gradually but steadily declined to a little over 16,000 in 1990. It seems clear that, at least for Air Force pilots, both inventory and requirements have been characterized by a steady, overall decrease, with some slight intermittent increases not affecting the overall downward trend from the late 1960s to the present time.

As seen in Table 3-2, at the end of the defense fiscal year 1994, Air Force pilot inventory stood at 14,634. Current projections of Air Force requirements (Thie et al., 1995:9-13) are for active duty pilot levels of 13,700 per year through the year 2002.

Table 3-2 shows that the pool of military maintenance personnel is also shrinking. For the period 1985-1994 the number of aviation maintenance officers declined at about the same rate as fixed-wing pilots, and enlisted aviation maintenance personnel have seen their ranks diminish more rapidly. The military has undertaken several initiatives to prepare military mechanics for transition to civilian jobs. For example, the FAA and the Department of Defense have developed a joint strategy aimed at converting military maintenance courses into credit for an airframe and power plant mechanic's license, thereby enabling military aircraft maintenance technicians to qualify for civilian aviation positions upon separation or retirement (Federal Aviation Administration, 1994).

Another way of looking at the effects of the military drawdown on the supply of trained aviation personnel is to look at what is happening to the annual intake into the military "pipeline." Table 3-3 shows the number of military personnel in their first year of service in selected occupational categories from 1980 to 1995. First-year pilots (including officers assigned to pilot training classes) for all services numbered 2,880 in 1980. That number ranged from roughly 3,500 to 3,900 throughout the 1980s, then began to fall sharply. In 1995 it stood at 1,494, or less than 40 percent of the total from 10 years earlier. The number of first-year maintenance officers was also lower in 1995 than in 1980, but with considerable fluctuations from year to year. The number of first-year enlisted aviation maintenance personnel has also varied greatly, but overall the 1990s have seen significantly fewer first-year enlisted maintenance personnel than the 1980s. The peak

TABLE 3-3 Active-Duty Personnel in Aviation-Related Occupations, First Year of Duty, All Services, 1980-1995

Occupation	1980	1981	1982	1983	1984	1985	1986
Officers							
Pilots	2,880	3,666	3,485	3,518	3,502	3,836	3,612
Fixed-wing pilots							
Fighter/Bomber	7	2	6	9	3	13	7
Other	58	216	78	28	4	10	29
Helicopter pilots	332	564	573	487	512	526	416
Pilot training	2,483	2,884	2,828	2,994	2,983	3,287	3,160
Aviation maintenance							
officers	343	303	350	376	435	259	237
Enlisted aviation							
maintenance personnel	9,886	12,711	9,004	5,851	7,577	10,057	10,060
Aircraft, general	4,618	6,922	3,995	2,747	3,510	5,366	5,528
Aircraft engines	1,788	2,073	1,275	1,177	1,336	1,529	1,694
Aircraft accessories	2,672	2,897	2,909	1,379	2,158	2,415	2,067
Aircraft structures	808	819	825	548	573	747	771

NOTE: Data shown are for the end of the defense fiscal year (September). Data for occupational categories correspond to the following occupational codes and designations: pilots (fixed wing, 2A and 2B; helicopter, 2C; pilot training, 9B/902); aviation maintenance officers (4D); maintenance enlisted personnel (600, 601, 602, 603) (U.S. Department of Defense, 1994).

intake for that category was 12,711 in 1981, dropping erratically from that point to a total of 3,218 in 1995.

Reductions in the numbers of pilots and enlisted maintenance personnel being trained in the military will eventually translate into fewer trained personnel who will be available to the civilian air carriers. Projections of the size of pilot training classes and military separations illustrate this point.

The Air Force, for example, expects smaller undergraduate pilot training classes for the foreseeable future. Whereas such classes averaged 2,200 from 1971 to 1980 and 1,850 from 1981 to 1990, they were targeted at 500 for fiscal years 1994 and 1995 and 800 for fiscal years 1996 and 1997 (Levy, 1995:36). Today's relatively small classes will translate into smaller numbers of military pilots becoming available to the civilian sector after the turn of the century.[1]

Military pilot separations from the services (Air Force, Navy, and Marines) had been running over 3,000 per year from 1990 to 1992 (Levy, 1995:31), providing a substantial pool of pilot candidates for the civilian air carriers. Projec-

---

[1] Air Force pilots can separate from the military and elect to fly for the airlines after completing their active-duty service obligation, generally after 10 years of service (Levy, 1995:30,36).

1987	1988	1989	1990	1991	1992	1993	1994	1995
3,268	3,205	3,518	3,383	2,344	2,095	2,123	1,252	1,494
15	3	22	8	3	4	1	0	18
19	20	57	47	23	35	7	6	5
343	340	391	399	327	350	708	322	260
2,891	2,842	3,048	2,929	1,991	1,706	1,407	924	1,211
198	256	248	91	191	164	212	275	244
7,939	6,949	5,630	3,714	4,153	3,567	3,516	4,519	3,218
3,960	2,903	2,813	1,673	1,875	1,604	1,975	2,261	1,696
1,141	964	773	431	727	641	352	626	383
2,003	2,199	1,363	776	963	927	833	1,238	792
835	883	681	834	588	395	356	394	347

SOURCE: Unpublished data provided by Defense Manpower Data Center, U.S. Department of Defense.

tions of future pilot losses show military pilot separations from the services decreasing to about 1,700 annually between 1995 and 1999, with a further decline to 1,500 annually by 2002 (Levy, 1995:31). In other words, the number of military-trained pilots who become civilians and perhaps join the air carrier applicant pool after the turn of the century will be half of what it was as recently as 1992.

THE MILITARY DRAWDOWN: IMPLICATIONS FOR MINORITIES AND WOMEN

A special focus of this report is access to civilian aviation careers by minorities and women. As we have seen, the military is a key source of trained aviation professionals. Equally important, it has been in the forefront of efforts to recruit and train minorities and women in a broad range of occupations, including aviation-related ones. (For a detailed discussion of women in the military, see Binkin, 1993; Binkin and Bach, 1977; and Holm, 1992. For a detailed discussion of blacks in the military, see Binkin, 1993; Binkin and Eitelberg, 1982.) One concern about military downsizing, therefore, is that it will reduce training opportunities for those underrepresented groups and restrict the diversity of the applicant pool for air carrier jobs. The impact of military downsizing on the diversity of the civilian applicant pool depends only in part on the size of the overall reduction in military trainees. It also depends on how many minorities and women have been receiving military training and how the drawdown affects these numbers.

Minorities in the Military

Since the end of conscription in the 1970s, the United States has created a military of "unparalleled diversity" (Binkin, 1993:1). Black participation in the military grew especially rapidly during that period. The proportion of blacks in the armed forces increased from 10 percent in the early 1970s to 20 percent by the beginning of the 1990s. In the Army, the growth was more dramatic, from 14 percent to 30 percent in the enlisted ranks (Binkin, 1993). This growth occurred against a historical backdrop of racial segregation and discrimination. Racial segregation of the military was formally ended in 1948 but persisted in policy and practice through the Korean War (Binkin and Eitelberg, 1982:26-30).

Whereas the overall participation rates of minorities, especially of black Americans, have been high, at least in the enlisted ranks, since the end of conscription and the transition to the all-volunteer force (Binkin and Eitelberg, 1982:Table 3-1), the representation of minorities is quite uneven across occupational specializations in both the enlisted and officer ranks (Binkin, 1993:78-79; U.S. President, 1995:43). Although there is evidence reflecting progress over time, there has also been a general recognition by the defense establishment that more needs to be done to enhance opportunities for minorities, especially in the officer corps (U.S. President, 1995:41).

To achieve greater opportunities and participation by minorities, the services have implemented a range of initiatives and programs. For example, the Army has developed the Enhanced Skills Program at historically black colleges and universities (combining mentoring and tutoring for students in academic trouble); the Army Preparatory School (preparing high school graduates for West Point with an extra year of academic study); and the Green to Gold Program (encouraging noncommissioned officers with at least two years of college to use the Montgomery G.I. Bill and Reserve Officer Training Corps (ROTC) scholarships to finish college and join the officer corps) (Kitfield, 1995:26). A key effort by the Navy and the Marine Corps has been the 12/12/5 Initiative, designed to ensure that officer and enlisted ranks across all ratings and specializations reflect the racial and ethnic diversity expected in the United States by the year 2000—12 percent black, 12 percent Hispanic, and 5 percent Asian-American/Pacific Islander (Kitfield, 1995:26; U.S. President, 1995:41). The Air Force, in cooperation with Delaware State College, a historically black college, has recently launched a summer flight training program targeting selected ROTC students from historically black colleges and universities around the country. This program, initiated in June 1996, is designed to give participating students the required skills to qualify for a private pilot's license and to enhance their chances of being selected for an Air Force pilot training slot upon college graduation.

Minorities are not as well represented in aviation-related military occupations as they are in the respective overall totals of either officers or enlisted personnel. The proportion of pilots who are black or Hispanic is especially small (Table 3-4). (We focus on blacks and Hispanics because we cannot separate other minorities from those whose ethnic status is unknown.)[2] In 1994, 10.1 percent of active-duty officers were black or Hispanic, but only 7.1 percent of all tactical officers and 3.0 percent of fixed-wing pilots were from these groups. In the enlisted ranks, 27.7 percent of all personnel and 21.7 percent of active-duty electrical/mechanical repair personnel were black or Hispanic, compared with only 17.9 percent of aircraft repairers.

Table 3-5 shows the numbers of minorities in the military in selected occupations for 1980, 1985, 1990, and 1994 and the percentage share of blacks and Hispanics in each occupation in these same years. Minority shares increased across the board between 1980 and 1994 (though the changes in the enlisted ranks were small). The percentage of blacks and Hispanics in each aviation-related occupation in 1994 was higher than or the same as it was in 1990, even though downsizing had reduced overall numbers.[3] The number of minority fixed-wing pilots and enlisted aircraft maintenance personnel dropped (slightly for the former but quite substantially for the latter). There were actually more black and Hispanic officers in the aircraft crew and helicopter pilot specialties in 1994 than in 1990. Helicopter pilots have traditionally not been hired in large numbers by the civilian airlines, although this could change as the number of ex-military pilots trained in fixed-wing aircraft declines.

Women in the Military

Women have a special history and experience in the military that is distinct from that of minorities (Binkin and Bach, 1977; Binkin and Eitelberg, 1982; Binkin, 1993; Holm, 1992). Until recently they were specifically excluded from

---

[2]We note that the overall conclusions in this paragraph would have been the same, although the absolute numbers would have been larger, if we had assumed that the "unknown" number was negligible and made our calculations including the "other minority/ethnic status unknown" group in the minority totals.

[3]Committee members were concerned that minorities might leave the service at disproportionately high rates, but data on individuals separating from the military provided by the Defense Manpower Data Center indicate this is not the case. Statistics for 1980, 1985, 1990, and 1994 show that minorities in aviation-related military occupations left the service at rates generally proportional to their representation in the various aviation occupational inventories.

TABLE 3-4   Active-Duty Personnel in Aviation-Related Occupations, All Services, by Sex, Race, and Ethnic Status, 1994

Occupation	Total Male and Female	Total Black and Hispanic	Black	Hispanic
All officers	245,841	24,739	18,312	6,427
Tactical operations	95,707	6,802	4,667	2,135
Fixed-wing pilots	21,476	644	346	298
Army	413	27	20	7
Navy	4,966	173	74	99
Air Force	14,634	394	230	164
Marines	1,463	50	22	28
Helicopter pilots	16,410	924	551	373
Aircraft crews	10,758	604	335	269
Other operations	47,063	4,630	3,435	1,195
Engineering and maintenance	32,383	4,126	3,277	849
Aviation maintenance	6,412	544	409	135
Other occupations	117,751	13,811	10,368	3,443
Enlisted personnel	1,350,681	373,606	293,173	80,433
Electrical/mechanical repair	269,820	58,539	43,556	14,983
Aircraft and aircraft related	108,699	19,409	13,609	5,800
Other occupations	1,080,861	315,067	249,617	65,450

NOTE: Data shown are for the end of the defense fiscal year (September). Data for occupational categories correspond to the following occupational codes: tactical operations, 2; fixed-wing pilots, 2A, 2B; helicopter pilots, 2C; aircraft crews, 2D; engineering and maintenance, 4; aircraft mainte-

full participation in the armed forces, and their military career choices and opportunities have historically been strictly circumscribed.

Despite their active participation in a wide range of military occupations during World War II, including airplane mechanics, gunnery instructors, naval air navigators, and pilots, the number of women in the military declined from 266,000 to 14,000 between 1945 and 1948 (Binkin and Bach, 1977:7, 10). In addition to the reduction in overall numbers in the military services, women were also confronted by legislative and policy barriers that affected their access to certain military occupations and promotional opportunities. Women in the Navy and the Air Force were specifically excluded by the Women's Armed Services Integration Act of 1948 (68 Stat. 368, 373) from duty assignments in aircraft

Other Minority/ Ethnic Status Unknown	Total Female	Shares			Other Minority/ Ethnic Status Unknown	Total Female
		Total Black and Hispanic	Black	Hispanic		
9,149	31,831	10.1	7.4	2.6	3.7	12.9
2,603	2,364	7.1	4.9	2.2	2.7	2.5
357	315	3.0	1.6	1.4	1.7	1.5
11	7	6.5	4.8	1.7	2.7	1.7
53	71	3.5	1.5	2.0	1.1	1.4
269	237	2.7	1.6	1.1	1.8	1.6
24	0	3.4	1.5	1.9	1.6	0.0
411	442	5.6	3.4	2.3	2.5	2.7
293	177	5.6	3.1	2.5	2.7	1.6
1,542	1,430	9.8	7.3	2.5	3.3	3.0
1,363	3,147	12.7	10.1	2.6	4.2	9.7
242	590	8.5	6.4	2.1	3.8	9.2
5,183	26,320	11.7	8.8	2.9	4.4	22.4
67,088	163,196	27.7	21.7	6.0	5.0	12.1
14,285	13,810	21.7	16.1	5.6	5.3	5.1
5,058	5,307	17.9	12.5	5.3	4.7	4.9
52,803	149,386	29.1	23.1	6.1	4.9	13.8

nance, 4D; aviation maintenance (enlisted only), 600, 601, 602, and 603 (U.S. Department of Defense, 1994).

SOURCE: Unpublished data provided by Defense Manpower Data Center, U.S. Department of Defense.

engaged in combat missions. That legislation also circumscribed career opportunities since it provided that women could not serve in command positions or hold a permanent grade above lieutenant colonel (or commander in the Navy). The same act authorized the Army to establish the types of military duty to which women could be assigned. The services for some time interpreted and applied the term *combat* in ways that placed many military occupations off-limits to women (Binkin, 1993:11; Binkin and Bach, 1977:22).

Not until the adoption of the all-volunteer force in 1973 did the services begin to interpret *combat* in less restrictive terms (Binkin, 1993:11). The Air Force eventually began allowing women to fly tanker, surveillance, and cargo aircraft (Binkin, 1993:11). In the 1970s women also gained the opportunity for

TABLE 3-5 Active-Duty Personnel in Aviation-Related Occupations, All Services, By Sex and Minority Status, Selected Years

Category	1980	1985	1990	1994	Shares			
					1980	1985	1990	1994
All officers	278,014	309,137	296,886	245,841				
Female	21,467	30,321	34,241	31,831	7.7	9.8	11.5	12.9
Black and Hispanic	17,045	24,550	27,148	24,739	6.1	7.9	9.1	10.1
Tactical operations	86,722	116,037	121,110	95,707				
Female	661	1,992	2,461	2,364	0.8	1.7	2.0	2.5
Black and Hispanic	3,691	7,221	7,901	6,802	4.3	6.2	6.5	7.1
Fixed-wing pilot	25,597	26,862	24,591	21,476				
Female	71	329	372	315	0.3	1.2	1.5	1.5
Black and Hispanic	431	623	661	644	1.7	2.3	2.7	3.0
Helicopter pilot	12,629	20,276	18,690	16,410				
Female	31	339	407	442	0.2	1.7	2.2	2.7
Black and Hispanic	335	806	891	924	2.7	4.0	4.8	5.6
Aircraft crew	11,966	12,525	12,181	10,758				
Female	16	155	178	177	0.1	1.2	1.5	1.6
Black and Hispanic	445	552	581	604	3.7	4.4	4.8	5.6

Engineering and maintenance	44,574	48,325	39,548	32,383				
Female	2,021	3,268	3,631	3,147	4.5	6.8	9.2	9.7
Black and Hispanic	2,904	4,433	4,658	4,126	6.5	9.2	11.8	12.7
Aviation maintenance	8,028	8,069	7,388	6,412				
Female	401	558	615	590	5.0	6.9	8.3	9.2
Black and Hispanic	480	550	564	544	6.0	6.8	7.6	8.5
Enlisted personnel	1,758,658	1,828,282	1,732,414	1,350,681				
Female	148,771	179,049	188,913	163,196	8.5	9.8	10.9	12.1
Black and Hispanic	456,366	456,783	488,048	373,606	25.9	25.0	28.2	27.7
Electrical/mech. repair	348,461	368,245	347,765	269,820				
Female	11,647	14,418	16,178	13,810	3.3	3.9	4.7	5.1
Black and Hispanic	70,011	70,624	74,965	58,539	20.1	19.2	21.6	21.7
Aircraft maintenance	146,290	151,700	135,610	108,699				
Female	6,469	6,219	6,611	5,307	4.4	4.1	4.9	4.9
Black and Hispanic	24,441	25,779	24,084	19,409	16.7	17.0	17.8	17.9

NOTE: Data shown are for the end of the defense fiscal year (September). Data for occupational categories correspond to the following occupational codes: tactical operations, 2; fixed-wing pilots, 2A, 2B; aircraft crews, 2D; engineering and maintenance, 4; aviation maintenance, 4D; aircraft maintenance (enlisted only), 600, 601, 602, and 603) (U.S. Department of Defense, 1994).

SOURCE: Unpublished data provided by Defense Manpower Data Center, U.S. Department of Defense.

the first time to command aviation maintenance units (Binkin, 1993:13). Women were admitted to the Air Force Academy, West Point, and the Naval Academy for the first time in 1976, and the classes of 1980 became the first coeducational classes in the history of the academies (Holm, 1992: 305).

These policy changes were reflected in the numbers and occupational mix of women in the military (Binkin, 1993:10-12). Between 1970 and 1990, the proportion of women in the armed forces grew from less than 2 percent to more than 11 percent. In 1970, 90 percent of women in the military were assigned to so-called traditionally female occupations (medical and administrative), but by 1990 that had dropped to 50 percent. By 1990 more than half of all Army positions— 60 percent in the Navy and 97 percent in the Air Force—were considered gender neutral.

Other significant changes for women in the military were triggered beginning in 1991. In that year, the statutory ban on women serving as combat aviators in the Air Force and the Navy was lifted. Detailed questions about the extent of women's participation in so-called combat assignments remained unresolved, however. Those issues were finally addressed when the secretary of defense ordered changes in Department of Defense policy, directing the services to open more specialties and assignments to women, particularly aircraft engaged in combat missions (Aspin, 1993, 1994). As a result, additional opportunities have opened up for women, including pilot assignments. Very few pilot assignments in the Air Force and the Navy are currently off-limits to women (pilot assignments in the Marine Corps being the exception) (U.S. Department of Defense, 1994:3-3 to 3-11).

Given historical restrictions, it is not surprising that women constitute a small percentage of the aviation-related officer corps (Table 3-5). In 1994 women held 12.9 percent of all officer positions in the active duty forces; however, they represented 2.5 percent of the tactical operations officer group, which contains the principal aviation-related officer occupations. In 1994, women held 1.5 percent of fixed-wing pilot positions in all military services, 2.7 percent of all helicopter pilot positions, and 1.6 percent of all aircrew positions. They were somewhat better represented in the ranks of aviation maintenance officers, at 9.2 percent.

Comparing the 1994 figures to data for selected years back to 1980 (Table 3-5), women officers, as a percentage of all tactical officers, have increased their participation in the active-duty officer ranks and in aviation-specific occupations. For example, from 1980 to 1994, the proportion of women has increased for each of the following occupational groups: tactical officers from 0.8 percent to 2.5 percent; fixed-wing pilots from 0.3 percent to 1.5 percent; helicopter pilots from 0.2 percent to 2.7 percent; air crews, from 0.1 percent to 1.6 percent; engineering and maintenance officers, from 4.5 percent to 9.7 percent; and aircraft maintenance officers, from 5.0 percent to 9.2 percent. Although the proportion of female fixed-wing pilots remained unchanged between 1990 and 1994, their number shrank from 372 to 315 because of the drawdown. The proportion of enlisted aviation maintenance personnel who were women changed little (4.4 to 4.9 percent) from 1980 to 1994. These proportions were substantially lower than the overall percentages of women in the enlisted ranks.

Downsizing and Training Opportunities for Women and Minorities

The military drawdown, as Table 3-3 showed, is reducing the numbers of people entering the military ranks each year. Tables 3-6 and 3-7 illustrate the effect the drawdown is having on the entry of women and members of specific minority groups into pilot and mechanic slots, respectively. Figures 3-3 and 3-4 summarize these data for white men and women and minority men and women.

The numbers indicate that, by and large, minorities and women have not lost ground in terms of their proportionate representation as the drawdown has proceeded. In the pilot ranks, there have even been some gains, for white women in particular.

The drawdown does mean, however, that training opportunities in the military are fewer, for minorities and women as well as for white men. Only 50 black men, 45 Hispanic men, 84 white women, and 8 black or Hispanic women in their first year of active duty in 1995 were pilots or in pilot training, for example. The numbers were larger in the period just before the drawdown began. The reduction in training opportunities is especially stark for mechanics, for whom the annual intake is significantly smaller in the 1990s than it was in the 1980s.

The data also lead to the conclusion that, despite the progress the military has made in opening pilot opportunities to minorities and women, the numbers in these groups remain a small fraction of the entering military pilot pool. The number of black men entering pilot programs is noticeably fewer than either women or other minority men; there has only been one year between 1980 and 1995 in which the total of all minority women from all racial/ethnic groups has exceeded nine. It is important to remember that these numbers represent the people entering pilot programs in the military and thus represent an upper bound on the pilots eventually leaving the military and entering the rank of civilian pilots. The actual number of former military pilots available to civilian airlines will be smaller, since some pilots entering military training will not complete it and some pilots leaving the military will not wish to become civilian airline pilots.

Even without these considerations, and even under optimistic circumstances, it is clear that the military cannot be expected to make a dramatic contribution to increasing the numbers of blacks, women, and other minorities in the cockpits of civilian airlines. For example, if the number of black men entering military pilot training was increased to 100 per year (a figure never yet reached), if all completed the program successfully, if all sought employment with the airlines upon

TABLE 3-6 Pilots in First Year of Duty, All Services, by Sex, Race, and Ethnic Status, 1980-1995 (percentages in parentheses)

Year	Total Male and Female	Total Male	Total Female	Total Known Race/Ethnic Status	White Male	Black Male	Hispanic Male	Other Minority Male	White Female	Black Female	Hispanic Female	Other Minority Female
1980	2,880	2,801	79	2,669	2,460	47	27	56	76	0	1	2
	(100)	(97)	(3)	(100)	(92)	(2)	(1)	(2)	(3)	*	*	*
1981	3,666	3,556	110	3,453	3,194	53	40	57	105	2	2	0
	(100)	(97)	(3)	(100)	(92)	(2)	(1)	(2)	(3)	*	*	*
1982	3,485	3,384	101	3,295	3,030	74	40	57	90	3	0	1
	(100)	(97)	(3)	(100)	(92)	(2)	(1)	(2)	(3)	*	*	*
1983	3,518	3,429	89	3,270	3,009	68	50	65	76	2	0	0
	(100)	(97)	(3)	(100)	(92)	(2)	(1)	(2)	(2)	*	*	*
1984	3,502	3,397	105	3,473	3,174	89	52	54	98	2	4	0
	(100)	(97)	(3)	(100)	(91)	(3)	(1)	(2)	(3)	*	*	*
1985	3,836	3,733	103	3,777	3,477	80	60	59	93	0	6	2
	(100)	(97)	(3)	(100)	(92)	(2)	(2)	(2)	(2)	*	*	*
1986	3,612	3,492	120	3,556	3,220	71	68	77	113	2	2	3
	(100)	(97)	(3)	(100)	(91)	(2)	(2)	(2)	(3)	*	*	*
1987	3,268	3,155	113	3,259	2,931	67	75	73	104	2	1	6
	(100)	(97)	(3)	(100)	(90)	(2)	(2)	(2)	(3)	*	*	*

Year												
1988	3,205 (100)	3,101 (97)	104 (3)	3,174 (100)	2,888 (91)	54 (2)	47 (1)	81 (3)	100 (3)	1 *	1 *	2 *
1989	3,518 (100)	3,404 (97)	114 (3)	3,475 (100)	3,116 (90)	86 (2)	60 (2)	101 (3)	103 (3)	4 *	2 *	3 *
1990	3,383 (100)	3,247 (96)	136 (4)	3,371 (100)	3,008 (89)	91 (3)	59 (2)	77 (2)	131 (4)	4 *	1 *	0 *
1991	2,344 (100)	2,238 (95)	106 (5)	2,342 (100)	2,054 (88)	59 (3)	36 (2)	87 (4)	101 (4)	3 *	0 *	2 *
1992	2,095 (100)	1,994 (95)	101 (5)	2,086 (100)	1,824 (87)	54 (3)	42 (2)	65 (3)	92 (4)	3 *	1 *	5 *
1993	2,123 (100)	2,015 (95)	108 (5)	2,112 (100)	1,788 (85)	67 (3)	45 (2)	106 (5)	98 (5)	4 *	1 *	3 *
1994	1,252 (100)	1,164 (93)	88 (7)	1,238 (100)	1,011 (82)	53 (4)	36 (3)	50 (4)	81 (7)	6 *	0 *	1 *
1995	1,494 (100)	1,398 (94)	96 (6)	1,468 (100)	1,232 (84)	50 (3)	45 (3)	45 (3)	84 (6)	6 *	2 *	4 *

NOTE: Data for "other minority" include American Indians/Alaskan Natives, Asian/Pacific Islanders, and others. Data for pilots reflect the following occupational categories: fixed-wing (2A & 2B); helicopters (2C); and pilot training (9B/902) (U.S. Department of Defense, 1994).

*Less than 0.5% of total pilots.

SOURCE: Unpublished data provided by Defense Manpower Data Center, U.S. Department of Defense.

TABLE 3-7 Aircraft Maintenance Enlisted Personnel in First Year of Duty, All Services, by Sex, Race, and Ethnic Status 1980-1995 (percentages in parentheses)

Year	Total Male and Female	Total Male	Total Female	Total Known Race/Ethnic Status	White Male	Black Male	Hispanic Male	Other Minority Male	White Female	Black Female	Hispanic Female	Other Minority Female
1980	9,886	8,935	955	9,880	7,183	1,073	373	302	826	80	19	24
	(100)	(90)	(10)	(100)	(73)	(11)	(4)	(3)	(8)	(1)	*	*
1981	12,711	11,767	944	12,705	9,853	1,077	451	380	807	87	25	25
	(100)	(93)	(7)	(100)	(78)	(8)	(4)	(3)	(6)	(1)	*	*
1982	9,004	8,287	717	8,996	6,819	935	305	220	597	84	14	22
	(100)	(92)	(8)	(100)	(76)	(10)	(3)	(2)	(7)	(1)	*	*
1983	5,851	5,487	364	5,850	4,721	448	190	127	310	40	8	6
	(100)	(94)	(6)	(100)	(81)	(8)	(3)	(2)	(5)	(1)	*	*
1984	7,577	7,203	374	7,577	6,234	578	229	162	328	33	6	7
	(100)	(95)	(5)	(100)	(82)	(8)	(3)	(2)	(4)	*	*	*
1985	10,057	9,449	608	10,056	8,023	882	303	240	540	50	9	9
	(100)	(94)	(6)	(100)	(80)	(9)	(3)	(2)	(5)	*	*	*
1986	10,060	9,532	528	10,054	7,975	908	358	285	469	35	15	9
	(100)	(95)	(5)	(100)	(79)	(9)	(4)	(3)	(5)	*	*	*
1987	7,939	7,526	413	7,934	6,280	707	335	200	359	30	13	10
	(100)	(95)	(5)	(100)	(79)	(9)	(4)	(3)	(5)	*	*	*

Year												
1988	6,949	6,407	542	6,947	5,353	530	308	214	445	60	23	14
	(100)	(92)	(8)	(100)	(77)	(8)	(4)	(3)	(6)	(1)	(*)	*
1989	5,630	5,052	578	5,626	4,210	423	207	208	454	70	34	20
	(100)	(90)	(10)	(100)	(75)	(8)	(4)	(4)	(8)	(1)	(1)	*
1990	3,714	3,437	277	3,712	2,874	282	173	106	219	33	18	7
	(100)	(93)	(7)	(100)	(77)	(8)	(5)	(3)	(6)	(1)	(*)	*
1991	4,153	3,930	223	4,151	3,374	266	183	105	175	25	15	8
	(100)	(95)	(5)	(100)	(81)	(6)	(4)	(3)	(4)	(1)	(*)	*
1992	3,567	3,365	202	3,562	2,901	212	167	80	162	10	15	15
	(100)	(94)	(6)	(100)	(81)	(6)	(5)	(3)	(5)	(*)	(*)	*
1993	3,516	3,317	199	3,511	2,768	234	192	118	155	25	12	7
	(100)	(94)	(6)	(100)	(79)	(7)	(5)	(3)	(4)	(1)	(*)	*
1994	4,519	4,205	314	4,519	3,437	377	245	146	263	23	18	10
	(100)	(93)	(7)	(100)	(76)	(8)	(5)	(3)	(6)	(1)	(*)	*
1995	3,218	3,011	207	3,215	2,425	264	191	128	163	19	12	13
	(100)	(94)	(6)	(100)	(75)	(8)	(6)	(4)	(5)	(1)	(*)	*

NOTE: Data for "other minority" include American Indians/Alaskan Natives, Asian/Pacific Islanders, and other minorites. Data for aircraft maintenance enlisted personnel reflect occupational categories 600, 601, 602, and 603 (U.S. Department of Defense, 1994).

*Less than 0.5 percent total aircraft maintenance personnel.

SOURCE: Unpublished data provided by Defense Manpower Data Center, U.S. Department of Defense.

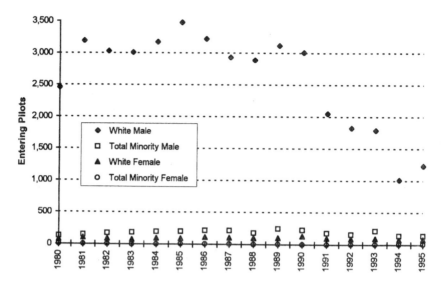

FIGURE 3-3 Pilots, first year of duty, all services, by sex, race, and ethnic status 1980-1995. SOURCE: Unpublished data provided by Defense Manpower Data Center, U.S. Department of Defense.

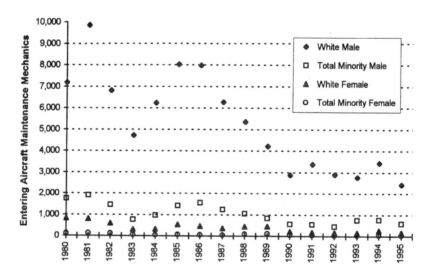

FIGURE 3-4 Aircraft maintenance enlisted personnel, first year of duty, all services, by sex, race, ethnic status, 1980-1995. SOURCE: Unpublished data provided by Defense Manpower Data Center, U.S. Department of Defense.

leaving the military, and if all were hired by the majors, they would still constitute only about 3 percent of the typical annual pilot hires by the major airlines. And if all were hired by the majors, there would be none for the smaller jet carriers or for the regional airlines. The story is similar for women and for other minorities. It seems clear, then, that for significant progress to be made in reducing the underrepresentation of minorities and women in the airlines, attention must be focused on the civilian training pathways to a pilot career. Furthermore, this would be true even without military downsizing.

4

Civilian Training for Aviation Careers

The downsizing of the military poses a new challenge for commercial aviation in the United States. Historically, the major airlines have relied heavily on pilots who received their initial training in the military. The smaller jet airlines, cargo airlines, and regional or commuter airlines have used both military-trained pilots and nonmilitary pilots trained in the large U.S. general aviation sector. Because of the availability of trained pilots from these two sources, U.S. airlines have not had to provide initial pilot training themselves. The military has also been a source for trained aircraft maintenance technicians. With continued military downsizing, the major airlines will have to turn increasingly to other sources for trained pilots and technicians. As the major airlines hire smaller numbers of military pilots and military mechanics, the effects are likely to ripple through other parts of the aviation industry.

There have been several periods in the past when the combination of high demand for pilots and a relatively smaller supply of military-trained pilots has caused the airlines to alter their hiring practices. Particularly telling is the most recent period of the late 1980s, when the initial phases of the current wave of military downsizing coincided with vigorous growth in commercial aviation. The airlines turned increasingly to civilian-trained pilots, and future shortages of pilots and mechanics were widely predicted.

This heavier reliance on civilian-trained job applicants drew greater attention to the differences between the structured and consistent backgrounds of individuals trained in the military and the diverse and more varied experiences of those trained through civilian aviation pathways. But before the issues surrounding civilian training could be explored fully, a sudden reversal of economic fortune subjected the airlines to a tumultuous period of recession, bankruptcy, consolidations, and shakeout. Such prominent companies in commercial aviation history as Pan American, Braniff, and Eastern disappeared. The industry lost billions of dollars in the early 1990s.[1] New hiring plummeted as thousands of experienced pilots and mechanics became available because of layoffs and furloughs. New training initiatives, no longer needed to ensure a quality applicant pool, were aborted or deemphasized.

The transformation in the military's role in pilot and mechanic training was only deferred, not avoided. The same issues that began to draw attention in the late 1980s are emerging again, as airlines are recovering economically and are again facing the pressures of expansion with an aging workforce that will sooner or later need to be replaced. The question remains whether a training system less dependent on the military over the long run can be expected to provide the air transportation industry with the numbers and kinds of workers necessary to operate efficiently and safely.

This chapter examines the assets and limitations of various civilian pathways for training pilots and aviation maintenance technicians (AMTs) and considers options for strengthening civilian training. Our findings on these issues, important in themselves, also set the stage for consideration in Chapter 5 of ways to improve the diversity of the aviation workforce so that it better reflects the nation's population.

TRAINING PATHWAYS FOR SPECIALIZED AVIATION OCCUPATIONS

Training for pilots and mechanics, the two primary specialty aviation occupations discussed in this report, is heavily influenced not only by the presence of the military but also by the popularity of recreational flying in the United States. Of an estimated 665,069 active pilot certificates held in 1993, almost 400,000 belonged to student or private pilots (Federal Aviation Administration, n.d.: Table 7.1). In 1993, 102,146 aircraft were in operation primarily for personal use, dwarfing the 7,297 aircraft operated by air carriers. Business, corporate, and other commercial use accounted for about 75,000 more airplanes (Federal Aviation Administration, n.d.: Tables 5.1 and 8.1). The United States is unique in the extent of its general aviation flying, in large part because of its high levels of personal income, low population densities in much of the country, and extensive infrastructure of general aviation airports. The diversity of aviation activities in

---

[1]Morrison and Winston (1995:30, note 24) report that "a commonly cited estimate of the extent of losses during 1990-93 for both domestic and international service is $12.8 billion (after tax and interest payments)." They indicate several reasons why this figure may be potentially misleading but agree that "by any measure the airlines did sustain considerable losses during the early 1990s."

the United States has helped spawn a wide array of training opportunities for individuals who wish to become pilots or mechanics.

The committee grouped the various training options into five fundamental routes, or training pathways, available to those seeking careers in specialized aviation occupations:

1. Military training,
2. Foreign hires,
3. On-the-job training,
4. Collegiate training, and
5. Ab initio training.

The first pathway, military training, was discussed in the previous chapter. The second pathway, foreign hires, is discussed below, but only briefly because it has not been an important source of pilots or aviation maintenance technicians in the United States. The next three pathways are discussed in somewhat greater detail.

Foreign Hires

In principle, U.S. airlines could hire pilots and mechanics who received their training from foreign airlines, foreign militaries, or other foreign sources. The market for pilots and mechanics appears to be becoming more international, and it is already common for U.S.-trained pilots to fly for foreign airlines. However, there is currently something of an oversupply of U.S.-trained pilots, and consequently U.S. airlines have shown little inclination to make foreign hires. In practice, it does not appear that this pathway is much utilized. In the near future, as Europe deregulates its airlines and rapidly developing countries like China build up their air transportation systems, there is likely to be more exporting of American-trained personnel than importing of foreign-trained specialists. In the longer term, conditions could change, and foreign-trained pilots and mechanics could be a source for U.S. airlines.

On-the-Job Training

Both pilots and maintenance technicians can currently earn the necessary FAA licenses and certification without attending a comprehensive formal aviation education program, by passing specific tests and fulfilling other requirements. A variety of proprietary schools offer short courses to help people prepare for these tests, or they can prepare on their own, often while working at aviation-related jobs that provide flight time or other experience required for certification. They are not required to have completed FAA-certified curricula or attended specific schools to take the licensing exams.

Pilots

Most commercial airlines in the United States require or highly prefer that their pilots have college (undergraduate) degrees. More than 90 percent of new hires have this credential at the larger airlines, according to FAPA surveys. There is no requirement that the degree be in an aviation-related field, however, so some college-trained pilots actually receive their flight-related training in the on-the-job training (OJT) pathway. On this pathway, would-be professional pilots work their way up through the system, earning certificates (beginning with the private pilot certificate) and ratings as they are able to pay for flight instruction and accumulating the flight hours required by the airlines in a variety of ways.

Based on a nationwide survey, *Flight Training* magazine (1996:53-76) reported that more than 1,700 schools teach people how to fly. All flight instructors have to be certificated by the FAA in accordance with Part 61 of the Federal Air Regulations (14 CFR 61). Flight schools that meet specified personnel, aircraft, facilities, curriculum, and other operating requirements can receive additional certification from the FAA under Part 141 of the Federal Air Regulations, but schools are not required to have this certification to teach flying. Nor are individuals required to attend a Part 141-certificated school in order to take the examinations for the pilot certificates and ratings described in Chapter 2, although it appears that a substantial majority, particularly those receiving advanced pilot certificates, do attend such schools (Blue Ribbon Panel, 1993:34). Part 141-certificated schools that have been granted examining authority by the FAA can recommend their graduates for some of the initial pilot certificates and ratings (except flight instructor certificates, airline transport pilot certificates and ratings, and jet type ratings) without the candidates having to take the FAA flight or written tests or both.

In 1994, there were 649 pilot schools in the United States certificated under Part 141 (FAA, 1994a). Some of these are collegiate institutions, and are described below. The majority are commercial operations specializing in flight training and perhaps offering other aviation services, such as fuel and maintenance at general aviation and commercial airports. While Part 141-certificated schools must keep detailed records on their students, there is generally little available information (aside from FAA regulations) that describes the training curricula or other aspects of the experience of students at noncollegiate institutions.[2]

Once they have obtained the basic licenses and ratings, individuals wishing to fly for the airlines must generally pursue an often long period of employment

---

[2]Each Part 141 school must have on file a training course outline; it must also provide a copy to the FAA. No one to the committee's knowledge has written about how these outlines vary from school to school or has attempted to provide comparative information to prospective students.

in flight-related jobs to accumulate the flight time necessary for advanced certification. To qualify for commercial and air transport pilot licenses, pilots have to augment their basic technical training with hundreds of additional hours spent flying increasingly sophisticated aircraft under a variety of challenging circumstances. They may work as flight instructors or in a variety of commercial ventures (dusting crops, flying pipeline and power line patrols, ferrying aircraft, towing banners, etc.) to earn a living while building flying experience. Much of this experience is likely to be with piston-engine aircraft. They may then move on to nonscheduled cargo or passenger operations, still in smaller aircraft. Eventually they may be hired as a first officer (copilot) for a scheduled regional airline, for which they may fly turboprop (turbine engine) planes. Some of the larger regional carriers may also provide their first experience with jet aircraft. After serving as copilot and then captain with a regional airline, pilots may move into the "right hand" (copilot's) seat, or in some cases the flight engineer's seat,[3] of a major carrier.

This pathway into a pilot's job with a major airline is highly idiosyncratic; many pilots drop out along the way, and others spend their entire flying careers without ever being hired by a major jet airline. Little information is available except the anecdotal kind about who pursues this route, what it costs in time or money, and how pilots trained in this way fare in the hiring process as they attempt to progress through the various stages.

A few freestanding aviation training organizations, such as FlightSafety International, Simuflight, Sierra Academy, and American Flyers, provide sophisticated flight training oriented to the professional air transport pilot. While primarily in the business of providing initial (post-hire), recurrent, and upgrade training for airline pilots and mechanics on a contract basis with employers, and training pilots for foreign airlines (again on a contract basis with the airlines), these companies do permit individuals without airline sponsorship to enroll and learn to fly. These companies may also offer so-called bridge or transition programs to qualify flight school graduates for airline jobs. (Another use of the term *transition training* refers to pilots upgrading or transitioning from one type of aircraft to another, but that is not what we are describing here).

Again, there is little information available on who pursues these options. The more advanced training offered by these companies, including transition training and training leading to aircraft type ratings, apparently became more important for would-be airline pilots after the airline labor market tightened up during the early 1990s and pilot supply greatly exceeded demand. Airlines were able to raise their minimum qualification levels, and individuals frequently found that they needed to obtain advanced training at their own expense in order to

---

[3]Flight engineers are the third crew members on airplanes that require three-person crews. This position is rapidly disappearing, as airlines replace older planes with newer aircraft that require only two-person crews.

receive serious consideration by the major carriers. This phenomenon affected pilots coming up through collegiate aviation programs as well and is discussed further in the section on the collegiate pathway.

Aviation Maintenance Technicians

The on-the-job training pathway appears to have been a more important route into airline employment for maintenance technicians than for pilots. Under current rules, maintenance technicians do not have to be certificated or have attended a certificated Part 147 school to perform certain jobs. Mechanics can qualify to take the FAA's certification examinations and meet other requirements for the airframe and powerplant (A&P) certificate by working under the guidance of certificated mechanics in repair shops, as long as certain conditions are met pertaining to their work experience. FAPA cites an FAA estimate that approximately 20 percent of all new A&P certificates are issued to individuals who received their training in this way (White, 1994:17). The major airlines, however, seem to prefer graduates of certificated programs, as do the regional airlines. The Blue Ribbon Panel indicated that a majority of new hires by major carriers (ranging from 40 to 80 percent, depending on the airline) were graduates of certificated Part 147 schools, according to a 1992 survey, and that 90 to 95 percent of new hires at regional airlines had graduated from these schools (Blue Ribbon Panel, 1993:18-19).[4]

Collegiate Training

Collegiate training, as indicated above, is already the major training route for certificated maintenance technicians and is likely to become an increasingly important pathway for pilots as military training opportunities decline. Four-year collegiate institutions offer training for both careers, although they train more pilots than maintenance technicians. Two-year colleges and postsecondary vocational schools are more likely to emphasize training of maintenance technicians. (Because there is substantial overlap, we do not try to divide this discussion neatly between the pilot and maintenance technician pathways but rather delineate the differences as we go along.) Collegiate programs may take people with no prior aviation experience and train them from the beginning. This pathway is distinct from ab initio training, in that college students (in aviation as in all other fields) are responsible for the costs of their courses of study.

A handful of today's collegiate aviation education programs trace their history to the early days of flight, but many more have their roots either in the Civilian Pilot Training Program that was set up in 1939 or in the wartime pilot

---

[4]The committee was not able to locate this survey, and the Blue Ribbon Panel report provided no details of how the survey was conducted, so we are unable to judge the reliability of this information.

training programs that operated on numerous college campuses (Kiteley, 1995:C45). A New Deal program, the Civilian Pilot Training Program was sponsored by the federal Civil Aeronautics Authority to help the aviation industry and bolster the national defense by providing trained pilots in the event of war. When war came, it was transformed into the Wartime Training Service and was taken over by the military. Using private aviation contractors as well as schools and colleges, the Civilian Pilot Training Program/Wartime Training Service trained over 125,000 pilots for the military until the Army Air Force effectively ended it in 1944, when Army Air Force training plus pilots returning from combat proved sufficient to handle military needs; the program was formally ended in 1946 (Pisano, 1993). Interest in collegiate aviation programs continued and grew, however, spurred by the growth of the aviation industry after the war and by the popularity of recreational flying.

Today approximately 300 collegiate institutions offer nonengineering aviation programs. Pinning down the exact number of programs and students involved in these programs is difficult because the available sources of data are incomplete and inconsistent.

The most extensive source of information comes from the University Aviation Association (UAA), a professional organization representing the nonengineering sector of collegiate aviation education. According to the Association's most recent guidebook (1994), 280 postsecondary institutions in the United States, Puerto Rico, and Canada (3 of the 280) offer nonengineering aviation programs. A survey it conducted in academic year 1992-1993 elicited responses from 205 of these institutions.[5] Of the respondents, 93 were four-year colleges and 112 were two-year institutions. Some offered only aviation courses, without a major or minor, but most reported offering degrees in aviation: masters (8), baccalaureate (80), associate (128), and other (77).

Table 4-1 indicates the types of programs available and the numbers of institutions that reported offering each type. Flight education is the most widely available option, followed by maintenance and management education. (Flight training is also offered under the aircraft systems management option of the airway science program, which we discuss later in this chapter.) The survey results indicate that most institutions only offer one or two aviation education options, rather than the full array of curricular alternatives.

The collegiate aviation programs identified by the UAA survey should be viewed as a lower bound rather than a full picture of aviation-related preparation in higher education institutions. The association's 1994 guidebook indicates that 75 institutions that did not reply to its 1992-1993 survey apparently do offer some kind of aviation education program. In addition, some students prepare for aviation careers by majoring in engineering, astrophysics, or other aviation-

---

[5]Results of the survey are unpublished and were made available to the committee by the UAA.

TABLE 4-1   Collegiate Institutions Offering Nonengineering Aviation Education Degree Programs, 1992-1993

Program	Number of Institutions Offering
Airway science	
Aircraft maintenance management	11
Aircraft systems management	25
Airway computer science	17
Airway electronic systems	11
Airway science management	29
Air traffic control	17
Aviation maintenance (excluding airway science)	63
Aviation management (excluding airway science)	63
Aviation studies	27
Avionics (excluding airway science)	27
Flight education (excluding airway science)	95
Other	18

SOURCE: Unpublished data provided by the University Aviation Association.

related fields rather than in aviation education per se. Schukert (1994) identified over 400 colleges and universities offering aviation/aerospace/flight courses and/ or programs of some kind. Neither the UAA nor the Schukert survey attempts to capture programs of study at noncollegiate schools (such as specialized nondegree career schools) that provide maintenance training. Finally, some college-educated individuals who aspire to pilot planes will obtain flight training without majoring in an aviation or aerospace field. For all these reasons, statistics from UAA on specialized aviation education programs—although the most complete picture we have of collegiate aviation—understate to some unknown but likely significant degree the availability of college-trained individuals with an interest in and the qualifications for flying commercial planes.

The FAA certificates maintenance training schools under Part 147 of the Federal Air Regulations (14 CFR 147); 193 institutions were so certificated in 1995 (Federal Aviation Administration, 1995). Many are community colleges, although four-year colleges, area vocational-technical schools, and other specialized career schools are also on the list. About 70 percent of these schools belong to the Aviation Technician Education Council, an association of Part 147 schools, the aviation industry, and government agencies.

Pending changes in AMT training (in particular the additional training required for the new AMT(T) certificate described in Chapter 2) would add between 700 and 800 hours to the 1,900 primary training hours required for the basic AMT certificate. The cost of advanced certification is most likely to be borne by individuals rather than by the company providing maintenance services. In fact, the advisory panel that proposed the new regulations sees the changes as reducing the training that maintenance organizations must provide. Most existing Part 147 schools are not equipped with the facilities or equipment needed to provide advanced training or, in the case of the two-year institutions that dominate provision of AMT training, are not expected to increase the teaching hours required for degrees and certificates. Instead of relying on these providers for advanced training, the advisory panel reportedly hopes that regional centers of excellence will emerge to provide this training. Fitzgerald identifies several schools that are preparing for such a role, including the Pittsburgh Institute of Aeronautics, the Trans World Technical Academy in Kansas City, Missouri, Fairmont State College in Bridgeport, West Virginia, and Metro Tech Aviation Center in Oklahoma City (Fitzgerald, n.d.:38-39).

The U.S. Department of Education's National Center for Education Statistics also collects information on institutions offering college degrees and certificates by program area, through its Integrated Postsecondary Education Data System (IPEDS) annual survey. IPEDS conducts a census of accredited two- and four-year colleges and collects data on a sample of technical and vocational postsecondary schools. The response rate to this survey is generally high (for example, 92.3 percent in 1989-1990), so nonresponse is not thought to be a significant source of error (National Center for Education Statistics, 1994:388-389). IPEDS includes some information on postsecondary institutions broken down according to a Classification of Instructional Programs. Although this classification has for many years included aviation programs, the definition of these programs led to some classification problems until a revision took effect in 1991-1992 that agency staff believe has provided more accurate breakdowns in the aviation-related categories. Like the UAA survey, IPEDS reports on aviation programs do not take account of individuals who may prepare for aviation careers by majoring in engineering, general business or management, or other fields.

Table 4-2 indicates the number of institutions reporting via IPEDS that they offered subbaccalaureate degrees or certificates or baccalaureate degrees in aviation fields of study in 1992-1993. The definition differences between the Classification of Instructional Programs categories used in IPEDS and the degree program categories reported by UAA make the two surveys difficult to compare. The numbers suggest, however, that neither survey is fully capturing the true number of aviation degree programs. In the case of UAA, not all institutions with aviation programs replied to its survey. In the case of IPEDS, nonresponse is less of an issue, but since respondents are usually from the central recordkeeping office rather than the aviation department, they may classify aviation degree recipients in nonaviation categories when they report to IPEDS. In any case, it is clear that the existing surveys of aviation programs do not provide a full portrait of aviation education in collegiate institutions and understate the training capabilities of the collegiate system.

For many of the same reasons, data on the number of students enrolling in collegiate aviation programs are incomplete and almost certainly understate ac-

TABLE 4-2   Collegiate Institutions Granting Aviation Degrees and
Certificates, 1992-1993

	Number of Institutions Offering	
Program Category	Subbaccalaureate certificates and degrees	Baccalaureate degrees
Aircraft mechanic/airframe	49	
Aircraft mechanic/powerplant	26	
Aviation systems/avionics	23	
Aviation/airway science	9	20
Aircraft pilot/navigator (professional)	62	19
Aviation management	31	31
Other aviation programs[a]	30	31

[a]Includes air traffic control, flight attendants, and "aviation workers, other."

SOURCE: U.S. Department of Education, National Center for Education Statistics, Integrated Postsecondary Education Data System, Completion Survey, 1992-1993.

tual enrollment.   In addition to the UAA surveys already discussed, the Aviation Technical Education Council has also surveyed enrollments in aviation maintenance technician programs (some of which are also included in the UAA surveys).   Its findings are clearly affected by low response rates.   The council received responses to its 1993 survey from only 33 percent of its membership, which includes somewhat less than three-quarters of the certificated maintenance schools.   The data on enrollments in aviation education programs in Table 4-3 and on enrollments and graduates of maintenance/avionics programs in Table 4-4 should also be thought of more as minima than as precise estimates.   Table 4-3 is notable for the high number of students (over 16,000) reported by UAA to be enrolled in flight programs in 1992-1993, including students enrolled in the aircraft systems management program under airway science.   The fact that the council's survey of just a portion of its membership (Table 4-4) shows roughly the same number of maintenance and avionics students as reported by UAA (something over 9,000, including avionics) suggests that the real number may be substantially higher.

According to IPEDS, the number of students receiving degrees or certificates in aviation-related fields is significantly smaller than the enrollments in these fields (Tables 4-5 and 4-6).   The differences are larger than one might expect after accounting for the fact that programs of study take several years to complete and that some students undoubtedly drop out along the way.[6]   The discrepancies

---

[6]This includes students who switch from flight to aviation management or other programs because they cannot afford the additional costs of flight training; such students sometimes decide to pursue flight training in less expensive noncollegiate flight schools.

TABLE 4-3   Enrollment in Collegiate, Nonengineering Aviation Education
Programs, 1992-1993

Program	Enrollment in Degree Programs
Airway science	
Aircraft maintenance management	284
Aircraft systems management	1,165
Airway computer science	223
Airway electronic systems	269
Airway science management	918
Air traffic control	736
Aviation maintenance (excluding airway science)	8,359
Aviation management (excluding airway science)	4,584
Aviation studies	6,515
Avionics (excluding airway science)	1,529
Flight education (excluding airway science)	14,941
Other	1,845

NOTE: U.S. Air Force Academy enrollment is not included in this table.

SOURCE: Unpublished data provided by the University Aviation Association.

TABLE 4-4   Aviation Maintenance Technician School Enrollments and
Graduates, Selected Years

	Enrollments		Graduates	
Year	Airframe and Powerplant	Avionics	Airframe and Powerplant	Avionics
1991	8,949	1,216	4,211	496
1992	8,099	1,049	4,069	500
1993	6,000	732	3,295	335

NOTE: Data reflect survey responses from 50 schools with Part 147 certification that have affiliation with the Aviation Technician Education Council.

SOURCE: Unpublished data provided by the Aviation Technician Education Council (ATEC).

are especially noticeable for flight programs. Although those who conduct the surveys have no explanation for the differences (having not examined each other's results and methods), we suspect again that they are partly attributable to variations between aviation departments and central recordkeeping units as to how they classify students by degree program. It also appears that students take aviation courses while not necessarily majoring in aviation. Further confusion emerges when one compares IPEDS data on degrees and certificates awarded in

TABLE 4-5   Subbaccalaureate Degrees and Certificates by Sex of Student and
Selected Program Category, 1989-1990 to 1992-1993

Program Category	Year	Number of Degrees	Men	Women
Aircraft mechanic/	1989-1990	1,088	1,062	26
airframe	1990-1991	902	866	36
	1991-1992	2,198	2,075	123
	1992-1993	2,148	2,030	118
Aircraft mechanic/	1989-1990	331	322	9
powerplant	1990-1991	805	780	25
	1991-1992	1,049	1,012	37
	1992-1993	1,022	989	33
Aviation systems and	1989-1990	N/A	N/A	N/A
avionics	1990-1991	N/A	N/A	N/A
	1991-1992	1,125	1,052	73
	1992-1993	1,375	1,296	79
Aviation and airway	1989-1990	186	162	24
science	1990-1991	211	180	31
	1991-1992	413	366	47
	1992-1993	405	364	41
Pilot and navigator	1989-1990	794	723	71
(professional)	1990-1991	819	718	101
	1991-1992	881	774	107
	1992-1993	723	657	66
Aviation management	1989-1990	329	270	59
	1990-1991	351	280	71
	1991-1992	184	151	33
	1992-1993	168	132	36

NOTE: The new Classification of Instructional Programs (CIP) was initiated in 1991-1992. Prior to that time, somewhat different classifications were used for aviation education programs. The data for prior years were aggregated for the table above to conform to the new CIP.

SOURCE: U.S. Department of Education, National Center for Education Statistics, Integrated Postsecondary Education Data System, unpublished Completion Surveys, 1989-1990 to 1992-1993.

maintenance (including avionics) with FAPA's assertion in its guide to aviation maintenance careers that 20,000 people graduate annually with A&P certificates, about 3,000 of whom also have associate or baccalaureate degrees (White, 1994:17).

The IPEDS data, which are available by sex but not by ethnic/racial status, also confirm the continuing underrepresentation of women in aviation programs,

TABLE 4-6   Bachelor's Degrees by Sex of Student and Selected Program Category, 1989-1990 to 1992-1993

Program Category	Year	Number of Degrees	Male	Female
Aviation and airway	1989-1990	773	680	93
science	1990-1991	873	780	93
	1991-1992	1,568	1,426	142
	1992-1993	1,560	1,388	172
Pilot and navigator	1989-1990	508	482	46
(professional)	1990-1991	634	570	64
	1991-1992	958	868	90
	1992-1993	854	787	67
Aviation management	1989-1990	829	721	108
	1990-1991	834	738	96
	1991-1992	780	693	87
	1992-1993	1,138	998	140

NOTE: The new CIP was initiated in 1991-1992. Prior to that time, somewhat different classifications were used for aviation education programs. The data in the table above and accompanying figures have been revised to conform to the new CIP classification.

SOURCE: U.S. Department of Education, National Center for Education Statistics, Integrated Postsecondary Education Data System, unpublished Completion Surveys, 1989-1990 to 1992-1993.

considering that women constitute a majority of all college students. The proportion of women receiving aviation degrees and certificates does appear to exceed their representation in the current aviation workforce.

Students who complete undergraduate pilot-training-plus degree programs at collegiate institutions generally have not been considered qualified candidates for employment by the airlines. The typical graduate has only about 250 hours of flight time. Like their OJT counterparts, these flight graduates normally must accumulate a great deal more flight time and experience before an air carrier will hire them. How much more appears to depend mostly on how deep the pool of potential pilot candidates is and what their qualifications are. Even in the boom years of the late 1980s, when airlines were hiring many new pilots and worries about shortages abounded, new-hire pilots on average had credentials greatly exceeding those of newly minted graduates of collegiate aviation programs. As best we can determine, minimum airline requirements hover around 1,500 hours of flight time and 250 hours of multiengine aircraft time. Major airlines can demand significantly more.

After finishing college, these graduates, like the OJT pilots described earlier, can spend years working their way up through commercial, nonairline flying jobs before they have the qualifications to secure an airline interview. Low demand for airline pilots during much of the turbulent 1990s may have lengthened this pathway, as competition intensified and applicants sought to boost their credentials. While such a change is difficult to document,[7] it is consistent with what we would expect when supply outruns demand. Also consistent with a situation of oversupply, some regional airlines have shifted training costs from themselves to pilots by requiring new hires to pay for their own initial post-hire training and aircraft type rating training (FAPA, 1992:22; Proctor, 1994:37).

In response to these labor market conditions, flight schools have developed some alternative programs that shorten the time it takes would-be airline pilots to build up flight time and credentials. Both the schools and airlines sometimes refer to these programs as ab initio programs, but this terminology results in some confusion with older, sharply focused ab initio programs run largely by foreign airlines to train their own professional pilots "from the beginning"; these latter programs are not combined with general higher education and, not incidentally, are usually paid for by the airline. For clarity's sake, we reserve the term *ab initio* for the programs discussed in the next section.

One way to shorten the time it takes for collegiate and OJT pilots to obtain airline jobs is to place more emphasis on airline needs in undergraduate programs and/or to create specialized courses designed to bridge the gap between the qualifications of the typical collegiate aviation program graduate (or OJT pilot with limited experience) and the selection requirements of the air carriers. The University of North Dakota operates one of the better-known programs focused on airline needs. The university has modified its Spectrum program, originally an ab initio program used largely by foreign airlines, by creating an option for bachelor of science students that emphasizes airline-oriented training in its flight program; it features crew-oriented flight training rather than just solo training, uses flight simulators, and emphasizes additional flight time and multiengine time (Hughes, 1989; Wallace, 1989; Wilhelmsen, 1995b). Not surprisingly, costs of such a program are higher than traditional collegiate flight training. Completion of such a program does not guarantee a job, so the university has added other options, such as a residency program in which the Spectrum graduate flies with a participating airline—usually a commuter carrier—for 4 to 6 months before the airline makes a hiring decision. Other schools have built variants of airline-oriented training into their undergraduate flight programs.

---

[7]When airlines describe the qualifications they seek, they tend to describe minimum criteria rather than the actual characteristics of new hires. Survey data, such as that collected by FAPA, are based on very small and nonrandomly selected samples of new hires, making it difficult to determine the accuracy of cross-year comparisons. Moreover, we were unable to obtain FAPA data on the certificates held by and other qualifications of newly hired pilots for 1989, the peak hiring year when contrasts with the tight labor market of the early 1990s may have been greatest.

Pilots appear to have a growing number of options to enroll in specialized flight training that will help them bridge the gap between initial qualifications and airline requirements. Flight training schools such as FlightSafety International and Comair Aviation Academy, which provide new-hire training for regional airlines, give pilots chosen for training a way to reduce the time spent building hours and awaiting interviews by pursuing airline-specific training for which they have already been prescreened, although they must pay for the training themselves (Proctor, 1994:37). A recent pilot career guide described a different approach used by Avtar, a Miami-based flight school that offers programs for several different levels of pilots (Wilhelmsen, 1995a). Pilots with commercial licenses and instrument and multiengine ratings can enroll in a basic program that allows them to build up multiengine and pilot-in-command time in a 10-passenger propeller plane. A more advanced program provides training and experience in turboprop planes while also building second-in-command or pilot-in-command time. A third program to train pilots in Boeing 727 jets gives the candidate 300 hours of experience in the aircraft along with 100 hours of pilot-in-command time. Avtar plans expansion into Boeing 737 and 747 training as well.

Ab Initio Training

Ab initio ("from the beginning") training prepares individuals with no flying experience to become pilots. In effect, this is a civilian, airline-sponsored version of what the U.S. military does, beginning with so-called no-time pilots and training them directly. Ab initio training is not generally used by U.S. carriers.

Airlines outside the United States have turned to ab initio training because they have been less able than their American counterparts to recruit military-trained pilots and because they have smaller and less developed general aviation sectors to provide civilian-trained pilots. Recreational flying is often limited in other countries by the high cost of fuel and aircraft, the paucity of airports, concerns over security, and heavy restrictions on travel (Garvey, 1992:69). The solution for many of these airlines is to produce pilots on their own, through ab initio training programs that take carefully selected candidates with no flying experience and put them through intensive pilot training courses designed specifically to meet the airlines' needs. A major carrier like Lufthansa of Germany produces copilots for its Boeing 737 jets in about two and a half years and about 250 hours of actual flight training (Glines, 1990; Warwick and Randall, 1993).

Much of the ab initio training carried out under the auspices of foreign carriers takes place at least in part in this country. The United States offers several advantages for carriers wishing to train their own pilots, including good training facilities (schools with classrooms, instructors, and in some cases dormitories, aircraft, and simulators); the infrastructure to support large-scale, sophisticated student activity (air traffic control, airports, landing aids, and ground service companies); and the fact that English is the international language used in air traffic control (Garvey, 1992:69). Good weather for flying in the American southwest also appealed to Lufthansa, an ab initio pioneer (Horne, 1989:80).

Lufthansa's ab initio training program goes back to the 1950s, when the carrier faced the challenge of developing pilots in the post-World War II days after its air force had been grounded (Glines, 1990:18). Beginning in 1967, Lufthansa began conducting its primary flight training through the Airline Training Center, then located in San Diego but soon moved to Arizona. In 1992 Lufthansa acquired the center, then known as Airline Training Center Arizona, when it was training students sponsored not just by Lufthansa, but also by All Nippon Airways of Japan, Air France, Iberia of Spain, Swissair, and EVA Air of Taiwan (Horne, 1989; Warwick and Randall, 1993). In 1990 Lufthansa's training program cost $153,500 per enrollee, of which the student was responsible for $13,500, which could be repaid once he or she began flying for the airline (Nelms, 1990:17).

FlightSafety International opened the FlightSafety Academy in 1966 to prepare professional pilots through ab initio programs; it used to train mostly self-sponsored individuals from the United States but now finds its student body dominated by foreign students sponsored by international carriers like Swissair and Tyrolean. A journalist noted that, over the years, the academy has trained students sent by carriers in France, Austria, Italy, Switzerland, Africa, Korea, Hong Kong, Luxembourg, Ireland, Saudi Arabia, England, and Belgium. Swissair was not only training its own pilots at the Academy but also overseeing pilot training for CTA, Balair, and Crossair (Garvey, 1992). Sierra Flight Academy in Oakland, California, had performed this type of training for 38 foreign airlines by 1993 (Warwick and Randall, 1993:36). International Air Service Company has trained pilots for Japan Air Lines in Napa, California, since 1970 (Wilkinson, 1991).

British Airways had its own ab initio flight academy at Prestwick, Scotland, which provided a 70-week training course for which the airline paid about $80,000 per graduate (Griffin, 1990:28-29). The school was closed when the carrier halted pilot recruitment during the economic downturn of the early 1990s. As British Airways began to hire pilots again after a five-year hiatus, it also began preparing for the future by recruiting cadets for ab initio training to be carried out through contracts with Oxford Air Training College and Cabair in England and Adelaide Aviation College in Australia (Morrocco, 1995).

True ab initio training programs, along the lines of these foreign airline programs, which are paid for mostly or entirely by the airlines themselves, have not taken root in the United States, despite a flurry of attention during the late-1980s boom, when hiring was brisk. That period saw an unusual level of interest in the ab initio approach on the part of some major U.S. carriers. Eastern, Northwest, and United were particularly active in developing programs to base flight training at selected colleges on airline standards, as ab initio programs do, although they neither paid for training nor guaranteed jobs to graduates (Nelms, 1988, 1990).

Northwest undertook the most extensive effort to build an ab initio program. Its training arm, Northwest Aerospace Training Corporation, joined with the University of North Dakota to design a new training activity that was built on a three-party agreement among the university, Northwest Airlines, and Northwest AirLink commuters. The training program was based on a $2 million task analysis study of the knowledge needed by Northwest and Northwest AirLink pilots and involved significant airline investment in new training equipment. Questions about control over the program and about the high costs facing students who ultimately might not be hired by the airline apparently led to an end of the full partnership approach (Glines, 1990). The program is now run mostly by the university, and much of the ab initio training has been provided for foreign airlines.

A Framework for Assessing Training Pathways

The five pathways described above represent the routes available for U.S. airlines to meet their needs for a workforce with highly specialized skills. To help structure its thinking about these pathways, the committee identified seven important dimensions along which pathways might conceivably be compared. These dimensions (which are illustrated in Table 4-7) include:

1. The cost of training to the airline,
2. The cost of training to the individual job candidate,
3. The "time to workforce" or how quickly a candidate can be prepared for employment,
4. The quality of the employee produced by the training,
5. Whether the pathway is more or less likely to contribute to workforce diversity,
6. How readily the pathway will accommodate technological change, and
7. How quickly the pathway can adjust supply in response to changes in the demand for employees.

The committee used the framework represented by the matrix in Table 4-7 to examine the likely advantages and disadvantages of each pathway for both would-be pilots and AMTs and for the airlines. The particular entries in the cells of Table 4-7 illustrate our thinking on how the various pathways compare in terms of preparing pilots for commercial airlines. These entries are not based on empirical study, because the necessary information is in many cases not available. Furthermore, within each pathway there is undoubtedly variation, sometimes substantial variation. The entries in each cell therefore represent the collective judgment of the committee about what different training options can offer U.S.

TABLE 4-7   A Framework for Assessing Training Options for Pilots

Dimension to Evaluate	Pathway					
	Military	Foreign Hire	On-the-Job Training	Collegiate	Ab Initio
Cost to airline	Low	Medium	Medium	Medium-low	High
Cost to individual	Low	Low	High	High	Low
Time to workforce	Medium-high	Medium-low	High	High	Low
Pilot quality	High	Highly variable	Medium-low, but highly variable	High	High
Changing workforce diversity	Potentially high	Low	Low	Medium	High
Technological adaptability	High	Medium	Low	High	Medium
Adaptability of supply	Low	Medium	Medium	Medium-high	High

carriers; they helped us assess which pathway or pathways are likely to dominate training for civilian aviation careers as the military diminishes in importance.

On the first dimension, the cost to the airline, it becomes clear even without detailed cost comparisons (which are not available) why the military pathway has been so attractive to the airlines and why ab initio training has not been adopted by U.S. airlines. From the airline's perspective, ab initio training paid for by carriers would involve the highest initial cost, whereas military training funded by taxpayers probably involves the lowest. Pilots who come to the airlines from the military already have received extensive training and experience with flying jet aircraft; the military's training programs are rigorous and demanding and weed out candidates that don't meet high standards. The attributes the military seeks are not completely the same as those of civilian airlines; the solo fighter pilot, for example, may be less attractive to some carriers than the transport pilot who has trained and flown in a more crew-oriented environment. Nevertheless, military training in general appears to serve a filtering or sorting-out role that the commercial companies value in their own selection processes. The other three pathways appear to entail costs to the airlines that generally lie somewhere in between military and ab initio training. Not much is known about what it would cost airlines to retrain foreign-trained pilots to operate in the U.S. flight environment, because this occurs so seldom. The foreign hire option could be somewhat more costly to the airline than the other two options depending on how difficult it turned out to be to familiarize foreign pilots with U.S. flight procedures and with cultural expectations that affect crew interaction in the cockpit. Collegiate programs structured closely to airline needs might require some industry financial support (for such things as equipment and curriculum development) but would still be less costly from carriers' perspective than absorbing all training costs themselves.

The second dimension, cost to the individual, of course produces a much different ranking. Here, the ab initio pathway (as it operates in foreign carriers) costs the individual the least. All or most of the training is paid for by the sponsoring airline and is designed specifically for its needs. Thus, the duration of the training is also the shortest, at least by comparison to the on-the-job training and collegiate pathways. The military pathway also offers the individual training with no out-of-pocket cost. However, the military pathway involves a substantial time commitment to the military, albeit with competitive pay and benefits, in exchange for the training. Moreover, military training is focused on military needs rather than on civilian airline needs. The on-the-job training and collegiate pathways involve the highest cost to the individuals, not only because the participants must pay for all of their own education and training, but also because they must typically work at low-paying jobs while gaining the additional certifications and flight time that the airlines desire for initial hires. In comparing the two, the collegiate pathway potentially has the edge in cost because, if closely adapted to airline needs, collegiate programs may result in less time being spent in nonairline flying jobs and therefore a shorter path to airline employment.

One problem with the more costly pathways is that individuals who cannot afford to pay for their training may be excluded from pursuing a pilot career. Although there is no reason to believe this cost problem would affect women more than men, it is likely, on average, to serve as a barrier to blacks and Hispanics more than whites, because blacks and Hispanics are more likely to have low-income backgrounds. Individual whites with low-income backgrounds would also be affected. These cost barriers are likely to increase in times when the applicant pool for pilot jobs is large relative to airline needs. In these circumstances, pilots are often asked to pay for an increasing share of their training, including getting the specific aircraft type ratings needed by the airline.

On the third dimension, time to the workforce, ab initio training, is clearly the quickest because it is focused exclusively on the needs of a sponsoring airline. If U.S. carriers attempted to hire foreign-trained pilots (assuming such a candidate pool were available), they might have to provide at least some additional training to equip their new recruits to operate according to the procedures and norms of American companies. Under the military option, pilots are trained to meet military needs relatively quickly, but these individuals become available to the civilian airlines only after their military service obligation is completed. Currently, the on-the-job training and collegiate pathways take a fairly long time. The time is highly variable, however, depending on the amount of experience required by the airlines and, especially in the case of on-the-job training, the financial capabilities of the would-be pilot. As with cost to the individual, the time to the workforce dimension has implications for participation by underrepresented groups. A pathway that takes a long time to the workforce and, as with on-the-job training, involves spending much of that time in low-paying jobs, may place a greater burden on individuals with low-income backgrounds than on those with higher-income backgrounds.

The fourth dimension, pilot quality, is difficult to assess. Surprisingly little objective information has been accumulated about operational flight performance. Job analyses describing the responsibilities of airline pilots are seriously outdated, and performance measures have been difficult to develop. Research has emphasized military rather than civilian pilot performance. Even here, most of the measurement effort has focused on fighter pilots, and little information is available concerning transport pilots (Damos, 1996:201-202). In the civilian sector, therefore, the concept of pilot quality is a rather subjective one, reflecting a sense of how satisfied the airlines are with the training and experience of pilots applying to them for jobs.

The preference that airlines have shown for military trained pilots reflects their belief that this pathway produces high-quality pilots. One would also expect that airlines undertaking ab initio training could design programs they believe produce high-quality pilots, particularly when they can require entrants to these programs to be college graduates (as the military does). Similarly, collegiate programs that are responsive to airline needs could produce high-quality pilots.

It may seem surprising to rank the ab initio and collegiate pathways similarly to the military. The military traditionally has provided very demanding training that quickly washes out those who cannot meet the rigorous standards. These latter two pathways would certainly vary from military training and would place different, and perhaps in some respects less rigorous, demands on pilot candidates. But because ab initio and collegiate programs can be designed specifically with the requirements of U.S. airlines in mind, they could produce pilots equally well suited (or better suited) to the demands of commercial flying as are military-trained pilots. For example, more emphasis might be placed on crew resource management and crew coordination in the civilian training pathways than in military training of single-seat jet fighter pilots. The foreign hire pathway probably would produce pilots of highly variable quality, in the view of U.S. employers. In part, this perception stems from statistics showing higher rates of fatal airline accidents and of fatal airline accidents with pilot error as the primary cause in regions of the world other than the United States and Canada (Oster et al., 1992:99). Even when foreign-trained pilots have received high quality ab initio training, both their training and their work experience are attuned to a flight environment outside the United States, with uncertain implications for their readiness to fly in this country. The on-the-job training pathway would also tend to produce pilots that vary greatly in quality, because their training and experience (most of it received in nonairline flying jobs) would not have been tailored to an airline flight environment.

In terms of the fifth dimension, changing workforce diversity, the ab initio and the military pathways offer the highest potential to improve the diversity of the workforce. Both of these pathways entail low cost to the individuals, so that financial capability is less of a potential barrier. Both have the capacity to undertake conscious, comprehensive efforts to improve diversity. But as seen in Chapter 3, the gains to date in the diversity of the military pilot workforce have been modest, and the military's service requirements mean that any improvements in the diversity of the military workforce are not reflected in more diverse applicants for civilian jobs for about 10 years. The collegiate program pathway could also make conscious efforts to improve diversity, but since it offers higher cost to the individual, financial barriers could prevent many low-income people from pursuing this pathway. The on-the-job training and the foreign hire pathways offer the least potential to improve workforce diversity. In the case of foreign hires, few foreign airlines employ substantial numbers of women or black pilots, so there simply isn't a pool to hire from. For the on-the-job training pathway, the cost to the individual is so high and the pathway so fragmented that it would be nearly impossible to put together coherent efforts to improve diversity.

For the sixth dimension, technological adaptability, both the military and collegiate pathways rank high. One of the strengths of military pilot training is the technological sophistication of the aircraft the pilots are trained to fly. Although civilian aircraft technology is different, military pilots are well trained to adapt to changing technological environments. The potential strength of the collegiate pathway is the foundation that a good basic education provides for adapting to new technological developments. The ab initio and foreign-hire pathways probably provide somewhat less potential for technological adaptability. Ab initio programs are likely to be aimed more at teaching effective use of current technology than providing a foundation for future technological change. Indeed, this very feature is part of the reason that they can achieve such quick time to the workforce. The foreign-hire pathway is much the same. The least-promising pathway in terms of technological adaptability is likely to be on-the-job training. Despite the sophistication of some smaller aircraft such as modern corporate jets, the jobs that pilots hold while they build flight time and experience on the job and pursue appropriate ratings generally involve less technologically sophisticated equipment than the airlines currently fly.

The seventh dimension, adaptability of supply to the fluctuating needs of the airline industry, has ratings similar to the cost to the airlines factor. Military training is the least adaptable to changing airline demand for pilots. First, the military bases its manpower needs on its mission, not on the needs of the civilian sector. Military and civilian needs are driven by quite different forces, and there is no reason to believe they will move in the same way. Second, even if the military were to respond to civilian pilot needs—and there is no reason why it should—the military training time and service commitments mean that there is a lag of 8 to 10 years between the time a candidate is accepted into military pilot training and the time that pilot is available for civilian service. At the other extreme, ab initio training has the potential to be most responsive to changing personnel needs because the airlines themselves control the programs and because it has the shortest time to the workforce. The remaining pathways have intermediate potential along this dimension. They would probably be more responsive than the military, because pilots working in other segments of the aviation industry, including foreign airlines, can be called up with less experience when major airline hiring demands are high, and they can be left in these segments to obtain more experience when airline demands are low. Flight training at collegiate aviation institutions could, with the guidance of the airlines, be modeled closely on airline requirements, mimicking to some extent the ab initio model that prepares pilots to enter commercial cockpits after comparatively short training periods.

In sum, this framework makes it clear why the military pathway, which combines low cost to the airline, rigorous selection and training procedures, and high technological adaptability, has been so attractive to the major airlines. But, as we have seen, the military's role as a provider of trained personnel for the airlines is clearly on the wane. Foreign-trained personnel do not appear to be much of a factor in U.S. aviation labor markets, perhaps because foreign airlines who often pay for training themselves are not likely to train large numbers of pilots beyond their own immediate needs. Nor has ab initio training been adopted much by U.S. airlines, most likely because its high cost does not seem justified as long as there is a substantial applicant pool of pilots trained in the military, collegiate programs, and on-the-job training.

This leaves on-the-job training and collegiate-based programs as the pathways that seem most likely to replace military training as the primary route to the major airlines. The committee expects that collegiate aviation increasingly will dominate on-the-job training because it can produce higher-quality pilots that are better able to adapt to changing technology. Collegiate aviation education, which embeds technical training in a broader foundation of learning, is potentially well suited to preparing workers who can adapt to new technology, learn continuously throughout their careers, and operate effectively in the team-oriented environment of the modern cockpit and maintenance workplace. As the experience of pilot candidates in the 1990s already proves, collegiate aviation programs, combined with the more advanced training offered by specialized flight schools, present individuals with numerous options for accumulating the extensive qualifications that airlines require when the labor market is tight.

In the committee's view, a civilian aviation training system grounded in collegiate aviation education offers the most practical alternative to the decline of the military pathway. The key question then becomes: Can a civilian training system that depends heavily on postsecondary education institutions to prepare aviation's most highly specialized workers be expected to provide the numbers and kind of workers that the air transportation industry will need to operate efficiently and safely?

CHALLENGES FOR A CIVILIAN TRAINING SYSTEM

With the downsizing of the military, it is clear that the civilian training system will have to play a larger role in meeting the pilot and AMT needs of the airlines. The committee concludes that the civilian training system, dominated by the collegiate pathway, can meet the specialized workforce needs of commercial aviation, both in terms of the number of people needed and the quality of the training provided. Before the system is fully effective, however, several challenges will have to be addressed.

Meeting Airline Demand

The initial question posed by military downsizing is whether civilian training will be able to make up numerically for reductions of military-trained aviation personnel, and the committee concludes that the answer is unequivocally yes. In

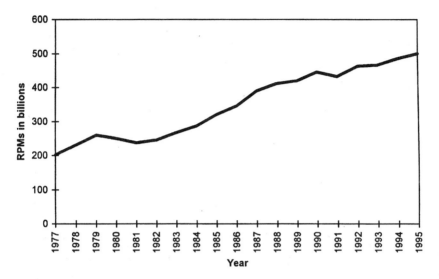

FIGURE 4-1 Airline industry system revenue passenger miles (Rpms). SOURCE: Form 41 reports submitted to the U.S. Department of Transportation by U.S. certificated air carriers, as compiled by Data Base Products, Inc.

reaching this conclusion, the committee did not construct a formal model of the supply and demand for specialized airline employment. Constructing such a model is difficult, and perhaps even futile, while the industry is undergoing such a fundamental restructuring.

Figures 4-1 and 4-2 show revenue passenger miles and enplanements for the U.S. airline industry system (domestic plus international) during the years 1977 through 1995. By both measures, the story is one of reasonably steady industry growth in the aggregate in passenger travel. There was a dip following the onset of the fuel crisis in 1979 and another more modest dip in 1991. To be sure, individual airlines did not all fare the same; some grew faster than others, a few entered bankruptcy, and others faced greater variation in traffic carried.

The first step in forecasting the need for airline employees is to forecast the future levels of airline travel on U.S. carriers. Airline travel depends on a variety of factors. Macroeconomic conditions, including interest rates, disposable income, growth in gross domestic product, and so on, are a major influence. And of course a major determinant is the price of airline travel, which in turn is strongly influenced by the price of fuel, interest rates, and other cost factors and by the degree and nature of competition, including the price of competing modes of short-haul air travel.

Predicting future air travel levels is difficult, but once that is done, it is even more difficult to predict the impact on employment because the relationship

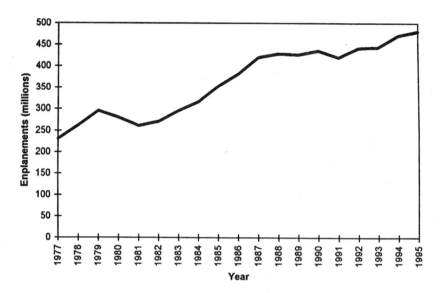

FIGURE 4-2 Airline industry enplanements. SOURCE: Form 41 reports submitted to the U.S. Department of Transportation by U.S. certificated air carriers, as compiled by Data Base Products, Inc.

between the number of employees, particularly pilots and mechanics, and the traffic carried has been changing as the industry restructures to meet the demands of domestic deregulation and the changing international aviation environment. One factor used to try to predict the need for pilots and mechanics is the projected number of aircraft needed to carry passenger demand. But the number of aircraft needed depends in part on the load factors, which measure the percentage of airline seats that are filled. As shown in Figure 4-3, load factors shot up immediately after deregulation as airlines used newly gained pricing freedoms to fill seats, then they dropped sharply with the onset of the fuel crisis in 1979. Since 1981, however, the increase in system load factors has been fairly steady. Indeed, the load factors have increased in all but four years, and following each of those four years they rebounded—and more—the following year. What will the load factors be in the coming years? Will they continue to rise? By how much and how quickly? Will they reach a certain level, then stabilize? What level and when? Or will the airlines intensify competition in a way that causes load factors to drop? When will they drop and how far? These are questions about which thoughtful and well-informed airline observers do not agree, but which are critical in forecasting the demand for pilots and mechanics.

Another critical issue is the average size of aircraft. One set of forces is pushing toward smaller aircraft, as airlines develop more point-to-point service overflying congested hubs. As air travel grows, however, there could be an

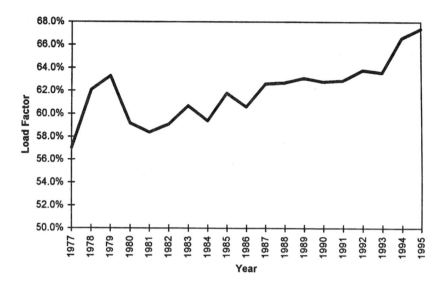

FIGURE 4-3 Airline industry system load factors. SOURCE: Form 41 reports submitted to the U.S. Department of Transportation by U.S. certificated air carriers, as compiled by Data Base Products, Inc.

opposing trend toward larger aircraft, as airports and airways become more congested and a growing shortage of new landing slots constrains the number of flights into key airports. Already the combination of local opposition and environmental concerns has made it increasingly difficult to expand airport capacity. These capacity constraints could be a particular problem in international service. Still another consideration is the aircraft crew requirement. Earlier generations of widely used aircraft, such as the Boeing 727 and DC-10, were configured for three-person flight deck crews, whereas the later generation of aircraft, such as the Boeing 757 and Boeing 767, were configured for two-person flight deck crews. These older aircraft can be reconfigured for two-person crews. Should that be done, the airline requirements for pilots could fairly quickly become lower than past practices would suggest. Even without this reconfiguring, these airplanes will eventually be phased out of service and are likely to be replaced with aircraft with two-person flight deck crews, again implying a lower pilot requirement than one might expect from past relationships between airline traffic and pilot requirements.

Taken together and subject to much uncertainty, these various potential developments suggest continued improvement in pilot productivity. The committee judges that there will probably be a continued upward drift in aircraft load factors. Despite countervailing forces, it seems likely that there will also be some increase in average aircraft size, as airport congestion limits growth in the number of flights at key airports. Average trip speed may also increase, as air traffic control procedures are improved and aircraft are allowed to fly more direct routes. Finally, there is likely to be a further downward trend in the average size of flight deck crews, although the rate of change is difficult to predict.

One might also expect improved productivity for AMTs in aircraft maintenance and repair. Changing aircraft technology could reduce the number of AMTs needed to support a given amount of passenger traffic. Also, growth in the use of third-party aircraft maintenance services could change the pattern of employment of AMTs and may contribute to greater productivity through greater specialization.

For these and other reasons, the committee did not base its assessment of the ability of the civilian training sector to meet airline needs on models of supply and demand. Rather, we based it on the demonstrated ability of the training sector to adapt to changing needs. In this regard, the civilian training sector may find the future easier to deal with than the past. In the past, the airlines have shown a strong tendency to hire military-trained pilots first and to hire civilian-trained pilots only as the stock of available military pilots became depleted. Thus, the volatile demand for civilian-trained pilots was linked not only to variation in airline hiring needs but also to variation in the numbers of pilots leaving the military. In the future, with the number of military pilots substantially reduced, the civilian training sector should face a larger and less volatile demand.

A second reason why we believe that the aviation labor market will adjust to airline needs is that the airlines are in a powerful position to influence this market. A study for the Department of Transportation succinctly stated the situation as it affects pilots (and, we would argue, maintenance personnel as well): "The supply of pilots is, to a large extent, within the control of the carriers—because they can provide the training to qualify new hires for pilot positions" (U.S. Department of Transportation, 1992:40). At the extreme, U.S. airlines would have the option already used by so many of their foreign counterparts, to provide ab initio training for carefully selected but inexperienced job candidates. But nothing in the U.S. experience suggests that this sort of ab initio training policy will be necessary or likely. U.S. carriers have many intermediate options for influencing the number and quality of the candidates available to them, short of undertaking and paying for training themselves. Their actions during the hiring boom of the late 1980s, when shortages were widely feared, are harbingers of the steps they can and undoubtedly will take if they are faced again with the possibility of not having all the qualified candidates they seek. And it is worth noting that, even in 1993, when a serious downturn made prospects for airline employment seem particularly poor, the FAA granted 12,645 new commercial licenses, 6,126 new ATP licenses, and 18,401 new mechanic licenses (Federal Aviation Adminstration, n.d.: Table 7.17).

Throughout their history, U.S. airlines have not needed to be heavily involved in developing an adequate labor pool because sufficient acceptable workers were available. The two partial exceptions occurred during the hiring booms of the mid-1960s and the late 1980s. During the first boom, airlines came as close as they ever have in this country to assuming responsibility for ab initio training, when some hired civilian pilots with only a private pilot certificate—although, at least at United, employment was apparently contingent on the pilot's receiving a self-funded commercial license (Blue Ribbon Panel, 1993:17; Warwick and Randall, 1993:36). Carriers were soon able to return to their traditional dependence on military-trained pilots, however. Despite the existence of numerous collegiate aviation education programs, there were few examples of industry-education partnerships, nor did the industry show much interest in influencing the direction of collegiate aviation programs. Not surprisingly, when carriers did begin to look at collegiate aviation to help avert the shortages predicted during the 1980s boom, they often found that collegiate programs offered only introductory courses, relied on old technology and equipment in their classrooms and laboratories, and had few faculty with aviation industry experience.

In the future, however, the prospect of a long-term reduction in the availability of military trained pilots is likely to change the role of collegiate programs and their relationships with the airlines. Numerous initiatives undertaken in the late 1980s indicate the potential for more productive industry-education collaboration. A spate of magazine articles describing these initiatives (e.g., Glines, 1990; Nelms, 1988; Parke, 1990; *Aviation Week & Space Technology*, 1989; Wallace, 1989), as well as a congressional hearing (U.S. Congress, 1989), testified to the level of concern about shortages and to the responsiveness of both airlines and schools. Northwest worked with the University of North Dakota to create the ab initio program described above. United forged relationships with Southern Illinois University at Carbondale and other schools (now totaling 20) to provide internships for advanced undergraduate aviation students at United's Denver training facilities; participants receive airline-oriented training, and some successful completers have interviewed with United for second officer (flight engineer) positions.

Before Eastern ran into the financial difficulties that eventually led to its bankruptcy and disappearance, the company had begun to forge relationships with two-year colleges to jointly prepare flight candidates; Eastern advised on the development of programs geared to airline requirements, on curriculum, and on flight standards. The airline also provided faculty for classroom instruction and was considering expanding its school involvement to aviation maintenance and avionics. Mesa Airlines developed a flight program in cooperation with San Juan College in New Mexico that in five semesters (including a summer session) was designed to provide students with an associate's degree in aviation technology, up to 300 hours of flight time, a commercial license with instrument and multiengine ratings, and (for successful graduates) a guaranteed job interview. Mesa agreed to waive its 1,500 total flight time requirement for program graduates who performed to airline standards. Embry-Riddle Aeronautical University, drawing on advice from industry, undertook a major restructuring of its flight degree program to give heavier emphasis to airline flight crew techniques, cockpit resource management, flight deck operations, and airline pilot proficiency development. These changes involved a significant investment in new simulators and training devices, computer-based instructional equipment, and a new training facility. Parks College, with guidance from its industry-based Career Advisory Board and with particular input from Delta, implemented a three-course avionics sequence as part of its maintenance management program.

Both airlines and schools have demonstrated in a variety of other ways that they could develop new approaches to averting shortages and other problems, such as rapid turnover at regional carriers, that accompanied a tight labor market. Continental, Northwest, and Pan American worked out partnership arrangements with regional affiliates to define career paths that would give the pilot a more structured advancement route from the regional to the major carrier while providing the regional carrier some stability in its workforce. Continental's plan, for example, involved hiring pilots at three different levels of experience, with pilots at the top level (who had 2,000 hours of flight time) destined for the cockpit at the major carrier. Pilots with 1,200 hours of flight time would be assigned to regional carriers owned by Continental to gain experience, earning a seniority number with Continental after 29 weeks. Pilots with about 500 hours of flight time would be sent by Continental through a ground school and simulator program estimated to cost about $8,000 and would then be assigned to a regional carrier, with their progress closely monitored. Even these so-called interns would get a Continental seniority number after 58 weeks.

TWA and FlightSafety International teamed up to develop a 30-day, $13,000 intensive transition course for pilots wanting to pursue jobs with regional carriers; this program featured loans from an independent loan company to help individuals finance the costs. Comair, a regional carrier based in Cincinnati, bought the Airline Aviation Academy in Florida to provide a 6- to 8-month, $21,000 initial pilot training program whose graduates would be prepared to enter an airline transition program. It too arranged loans through a credit company for its students. Air Carrier International Flight Academy used military and ab initio models to develop a 16-month, $51,500 program based on United Airline's requirements and designed to prepare both zero-time and experienced pilots for regional airline jobs.

We do not have any systematic information about the outcomes of these various innovations. At least some of them were aborted altogether when the economic fortunes of the industry reversed, and others were undoubtedly cut back or (like the University of North Dakota program) never fulfilled the mission initially envisioned for them, although they continue to exist. What these examples do show is that the airline industry can do much to influence the supply of workers, should shortages develop, and that the system has much capacity to respond.

Moreover, there is little sentiment today that shortages are likely to be a serious problem. Those who showed the most concern about the imbalances between supply and demand in the late 1980s take a different view now. For example, in 1990 FAPA estimated that there would be a worldwide airline pilot shortage of as many as 51,000 by the end of the century (Glines, 1990:18). In 1993 FAPA's president was quoted as saying that the organization "does not believe there will be a pilot shortage—ever" (Warwick and Randall, 1993:11). The Blue Ribbon Panel established to address congressional concern over the shortage fears of the late 1980s determined, by the time it concluded its study in 1993 (p. xiii), that:

> In the near term (3 to 5 years), the existing pool of experienced pilots and AMTs may be adequate to supply the needs of the air transportation industry. In the long term, there will continue to be adequate *numbers* of pilots and AMTs meeting minimum qualifications for their jobs in air transportation, but it is unlikely that enough of these pilots and AMTs will have the proper *skills and experience* to provide industry with sufficient numbers of *well qualified* personnel.

As this conclusion suggests, a second important question confronting civilian aviation is whether the airline industry—accustomed to the luxury of having much of its specialized workforce arrive with comparatively high-level skills—will have to take a more active hand to support the training it wants in an environment more heavily dependent on civilian (and particularly collegiate) training. The committee's view is that it will, and a key challenge for the future is professionalizing and standardizing civilian training, for the benefit of both the airlines and their potential employees.

Professionalizing and Standardizing Training

In 1990 the chairman of the aviation department at a major state university observed that collegiate aviation education was at the stage of "maturing from adolescence into young adulthood as a recognized and accepted curriculum on our nation's campuses" (letter from Stacy Weislogel of Ohio State University to Gary Kiteley of UAA, quoted in University Aviation Association, 1990:58). Largely ignored by industry and guided primarily by basic FAA regulations governing schools that offered training for flight and maintenance technician certification, collegiate aviation programs have developed in widely disparate ways for many years. Most programs were not explicitly geared to producing graduates ready for employment in commercial aviation. Airlines and other employers had little way of knowing, other than through firsthand acquaintance with particular schools, what specific training aviation graduates had received. Students had little way of knowing which schools' curricula were most likely to lead them to the jobs they wanted.

This situation has improved since UAA made its first attempt in 1976 to establish standards for curricula, courses, and credits, but much remains to be done to guide collegiate aviation from young adulthood to full maturity. From all we have been able to learn, the industry is still only minimally involved in helping the schools mature. As we have seen, promising school-industry partnerships were short-circuited by the economic reverses of the early 1990s. Wide disparities and comparatively little information about course content and standards still characterize collegiate aviation programs.

These disparities and information gaps grow more important as the pace of technological development accelerates in commercial aviation. Training needs are changing rapidly. Pilots, mechanics, and other employees will work in an environment increasingly characterized by highly sophisticated engines, electronic instrumentation ("glass cockpits"), fly by wire (computer-driven controls), computer monitoring and testing for all systems, and global positioning system and enhanced vision systems navigation equipment linked by flight management systems. Pilots and mechanics trained on aircraft using earlier technologies will require costly upgrade training. Facilities and instructors in schools providing pre-hire initial training often lag behind the current state of the art by 10 to 15 years. Crew resource management (how to operate as part of a crew, how to communicate as a team) is increasingly important to airlines, but flight training has typically emphasized preparing the solo pilot for the necessary FAA certificates rather than learning to fly as part of a team (Blue Ribbon Panel, 1993:23-26).

In such an environment, program disparities and information gaps are not in the interest of either the commercial aviation industry, which will probably have to rely more and more on collegiate aviation, or of the students, who need the best information possible to prepare effectively to work in an industry characterized by volatility and employment uncertainty. Important progress has been made by the schools themselves in reorienting their programs toward the needs of the modern aviation industry; the federal government has also aided this process through its sponsorship of the airway science program. Industry has been the rather noticeably minor partner. In the committee's view, it is no longer appropriate for industry to assume that the job of preparing its workforce belongs to someone else. It is time for aviation employers to take more responsibility for supporting the education and training system they need.

Some employers, as we have indicated, already help schools develop their programs: for example, by serving on curriculum advisory committees, assisting research efforts such as the AMT job task analysis and curriculum reform project currently under way at Northwestern University, providing internships and other opportunities for students, donating equipment, providing student financial aid, and other means. Such employer-school partnerships appear underdeveloped to us compared with such fields as engineering and computer science, and their further development should be encouraged. Industry must go beyond individual school relationships, however, and work more actively to help develop the field as a whole. Otherwise, industry will remain uninformed about what it is getting from many schools; colleges will be subject to inefficiencies resulting from a mismatch between what they offer and what their students need for success in the job market; and students will not have the information they need to make good educational choices.

The federal government has already given aviation a boost toward professionalizing and standardizing collegiate education and training programs through the airway science (AWS) program. In 1978 the FAA became interested in designing specialized postsecondary programs to educate its own future workforce. In 1983 FAA received approval from the federal Office of Personnel Management to initiate an Airway Science Demonstration Project. Under the project, FAA would develop model curricula for the preparation of air traffic controllers, electronic technicians, aviation safety inspectors, and computer specialists. Students who completed approved airway science programs would be ranked on and selected from a separate personnel register from the ones then in use. The demonstration project was designed "to compare performance, job attitudes, and perceived potential for supervisory positions of individuals recruited for several of FAA's technical occupations who have an aviation-related college-level education, or its equivalent, with individuals recruited for the same occupations through traditional methods" (Federal Register, Vol. 48, No. 137, July 15, 1983).

In cooperation with the UAA, the FAA developed five airway science programs of study for four-year colleges:

• Airway science management—preparation for a variety of aviation-related administrative and management positions in fields including air traffic control.

• Airway computer science—preparation for jobs involving computer operations, software design, systems analysis, and computer programming.

• Aircraft systems management—preparation for aircraft flight operations and preparation of professional pilots and flight instructors.

• Airway electronics systems—preparation for work involving the maintenance, troubleshooting, testing, and development of avionics and navigational equipment.

• Aviation maintenance management—preparation for work in maintenance and troubleshooting and for technical management roles.

In addition, in February 1993, the FAA initiated a two-year airway science program with three areas of specialization:

• Flight technology—preparation for aircraft flight operations and initiation preparation of professional pilots and flight instructors.

- Airway electronics technology—preparation for work in the maintenance, troubleshooting, testing, and development of aircraft avionics equipment or ground communication/navigation equipment.
- Aviation maintenance technology—provision of theoretical and practical knowledge pertinent to airframe and power plant maintenance and preparation in technical documentation methods, specifications, and standards.

Although the origins of the airway science program predate the air traffic controllers' strike of 1981, which resulted in the firing of most of the controller workforce, the strike no doubt increased interest in the program. Within the first five years (by 1988), 34 colleges had one or more FAA-recognized airway science programs (University Aviation Association, 1990:Appendix C). At the end of 1995, 60 four-year colleges and community/technical colleges had approved airway science programs.

Congress, too, became interested in the program and soon after its inception started appropriating funds for airway science grants. Between fiscal 1982 and fiscal 1993, over $104 million was provided to institutions of higher education for AWS buildings and equipment. Of this total, nearly $100 million was appropriated by Congress and earmarked for specific institutions. The remaining $4 million came from other FAA funds and was awarded through a competitive application process. The funds were distributed very unevenly among institutions: some received nothing, and five schools got more than half of the money (U.S. Department of Transportation, 1993).

In terms of its goal of contributing to the development of FAA's own workforce, the airway science program has been judged a failure. Both FAA-sponsored evaluations (FAA, 1990) and an audit by the Department of Transportation's inspector general (U.S. Department of Transportation, 1993) found that almost no graduates of airway science programs had been hired by the FAA. The FAA estimated in 1990 that 60 percent of airway science graduates had gone to work for the airlines, 35 had taken other aviation-related positions, and 5 percent went to the FAA or took nonaviation-related jobs. According to the inspector general, the dismal track record of the FAA in hiring airway science graduates resulted from several factors: three of the five airway science options lacked curricula fit to provide a hiring path to FAA positions; the FAA could not compete on starting salaries with private industry; and the lengthy FAA recruitment process compared unfavorably with other employers. The FAA itself acknowledged institutional resistance within the agency to hiring airway science graduates (Federal Aviation Administration, 1990).

There was disagreement between the FAA and the inspector general about whether the airway science program had been intended to benefit the industry as well as the FAA (the FAA argued that it had); whether intended or unintended, the FAA believed there were benefits to the civilian training system that prepared graduates for industry jobs. Its Office of Training and Higher Education cited several:

• Establishment of a strong aviation curriculum.
• Enhanced professionalism of aviation courses at the collegiate level, including the greater visibility and credibility provided by official FAA recognition.
• Development of accreditation standards: the experience of the University Aviation Association in developing airway science curricula, evaluating proposed curricula, and inspecting airway science programs encouraged the association to establish the Council on Aviation Accreditation in 1988 to ensure program quality and help improve the broader universe of aviation education programs.
• Grant support for buildings and facilities.

Of special interest to the committee, given our charge, is that the airway science program expanded the involvement in aviation education of historically black colleges and universities and of institutions serving large numbers of Hispanics. Historically black colleges and universities in particular have played a central role in the United States in producing black college graduates; they have been especially important sources of black scientists and engineers (Trent and Hill, 1994). The effort to involve these institutions in airway science has been cited as a model for attracting minorities to the transportation profession (including railroads, shipping, trucking, and highways), in which they have been underrepresented and largely excluded from management and professional positions (Ayele, 1991). At the end of 1995, 14 predominantly black four-year colleges and 2 predominantly Hispanic institutions had FAA-recognized airway science programs. Many of these institutions did not have aviation programs before the airway science program. Many of them are public institutions, and the FAA's endorsement through airway science recognition was important to their ability to receive state support. Delaware, for example, home of Delaware State University, requires its programs to have accreditation or other external approval. Maryland has decided that the new airway science program at the University of Maryland/Eastern Shore will be the only aviation education program at a public college in the state. At least 7 of the 16 predominantly black and Hispanic airway science schools received grants. One (Florida Memorial) was among the five largest institutional recipients (expenditure data provided to the committee by the FAA).

The federal role in the airway science program is now coming to an end. Following the inspector general's recommendation, the FAA agreed in 1993 that it would seek no further funds for airway science grants and would discourage congressional efforts to appropriate grant funds. The Clinton administration's National Performance Review, headed by Vice President Gore, also called for termination of the grants, principally on the ground that "[m]any schools now offer high quality aviation training programs without support from FAA" (Gore, 1993:98). No funds were included for grants in either the fiscal 1995 or fiscal 1996 appropriation bills for the Department of Transportation.

In addition to grants, the other key component of the airway science program was the development and approval of curricula at individual colleges and universities. Since the inception of the program, the University Aviation Association, under contract with the FAA, reviewed aviation curricula proposals for institutions establishing new airway science programs and made recommendations for FAA approval. The FAA has developed a plan to shift responsibility for identifying and developing curricula for this program to the UAA and to the fledgling accreditation agency, the Council on Aviation Accreditation (U.S. General Accounting Office, 1994:304-305; Federal Aviation Administration, 1994b). The FAA does not plan to fund UAA activities related to airway science after 1996. It will not recognize new programs, and, after a transition period during which current programs are expected to seek Council accreditation, the FAA will drop its formal recognition.

Airlines and the aviation industry as a whole have an important stake in the continuing maturation of collegiate aviation education as a recognized and accepted curriculum tied to the needs of the commercial sector. The committee believes that it is time for companies and their trade associations to become more active and systematic partners in fostering this development. For reasons sketched out earlier in this chapter, collegiate aviation is the most promising alternative to the military to become the main source of trained aviation personnel and is certainly a less costly alternative for the airlines than the ab initio training on which many of their foreign counterparts must rely. Fully exploiting the advantages of this training pathway requires more than the limited school-industry collaboration that has characterized American aviation in the past.

There is, however, a potential "free rider" problem here. Individual companies may think that, at least in the near term, their interests can be served by letting others bear the effort and expense of supporting collegiate aviation programs. Such a view would be short-sighted. These companies may find themselves less able to tap into collegiate-based training centers when aviation labor markets enter tight periods; schools will understandably be more willing to work with those who have helped them develop their programs. Nevertheless, the temptation to ride for free may be strong, which suggests that the aviation industry ought to think more about training from a collective perspective, not just company by company.

Companies have another reason to work collaboratively as well as individually with collegiate aviation education: the desirability of developing commonly recognized training standards. Airlines have distinctive cultures and procedures, as reflected in company-specific training requirements concerning specific route systems, ways of dispatching people and planes, and the like. But there are also generic training requirements that can be used to define common standards for collegiate aviation programs that would meet the needs of multiple employers. This approach ought to be attractive to employers, who have resisted supporting ab initio training and similar efforts because of reluctance to commit themselves, perhaps years in advance, to hiring program graduates. Common standards would provide schools with the knowledge they need to build their curricula. Carried out through a mechanism that recognizes or certifies programs that meet the standards, they would give prospective students the assurance that their training suits the demands of industry.

Standard-setting and program recognition in American postsecondary education are normally accomplished through a voluntary and self-regulating system of accreditation. National or regional associations of colleges and schools accredit entire institutions. Commissions on accreditation established by national professional organizations conduct specialized accreditation of professional and occupational schools and programs. Specialized accreditation is well established in other fields, such as business, medicine, law, and engineering, but is just in its beginning stages in aviation education. Eight institutions had received accreditation from the Council on Aviation Accreditation for aviation education programs by June 1996.

The committee recommends that collegiate aviation programs support the development of a system of accreditation similar to that found in engineering and business. The accreditation system should be developed through the Council on Aviation Accreditation or a similar organization working in close cooperation with the airlines to ensure that curricula are responsive to their needs. It should link its standards to the best available research on competency and skill areas, such as the AMT job task analyses currently being conducted at Northwestern University, and should encourage systematic evaluation of institutional programs.

The committee further recommends that the commercial aviation industry support development of an accreditation system as well as provide more sustained and consistent support to individual aviation programs. Active involvement of employers is crucial if aviation accreditation is to accomplish its goal of establishing uniform educational quality standards that meet industry needs. Some employers are already assisting in these efforts, but more widespread participation is needed. To minimize the free-rider problem, employers may want to work at least in part through their major trade associations, such as the Air Transport Association and the Regional Airline Association for the airlines, to help develop and update accreditation standards consistent with the training they believe their future employees will need. Trade associations should also work actively with their members to increase their understanding of accreditation and its benefits.

Employers also have an important role to play in making accreditation effective by giving special consideration to hiring graduates of accredited programs and by encouraging aviation programs to seek accredited status. In engineering, for example, Boeing does this by giving the title of "engineer" (and the accompanying pay) only to graduates of accredited engineering programs.

In addition to helping aviation education develop widely accepted quality standards, aviation employers ought to help individual schools develop their programs (as some do now). The committee's view is that university/industry interaction is less well developed in aviation than in engineering and other professional fields. As some examples earlier in this chapter illustrate, this interaction has also tended to be sporadic: heavy in the rare times when labor markets were tight and largely nonexistent otherwise. Colleges and universities attempting to build training programs that depend partly on private contributions (both monetary and nonmonetary) must be able to count on this assistance through both good economic times and bad. In the past, companies may not have had reason to think much about how their actions affected collegiate aviation, given the supply of trained people from the military. Now their own self-interest argues for becoming more sensitive to how training in civilian colleges and universities can best be developed.

The committee recommends that the FAA facilitate school-industry cooperation and the development of an aviation accreditation system. As we have indicated, industry-school relations are not yet well developed in aviation, and accreditation is in its early stages. Thanks in part to historical precedent, the industry is accustomed to having the FAA involved in important aviation-related discussions. The agency's participation in efforts to build university-industry ties and to develop an effective standard-setting mechanism that goes beyond the minimal requirements of the Federal Aviation Regulations will help legitimize these efforts. The FAA's official mission statement calls for it to foster civil aeronautics and air commerce. Contributing to the development of a strong civilian aviation education system is clearly within this charge.

The committee further recommends that the FAA review its training and certification requirements to ensure that they support rather than hinder the efficient and effective preparation of aviation personnel. In the time available to us, the committee could not undertake a thorough review of FAA requirements that affect the training offered in collegiate aviation programs. We did, however, become aware of several issues concerning FAA rules.

For example, FAA regulations require various kinds of solo training for pilot certifications that may not be as important for airline pilots, who always fly in a crew environment, as they are for recreational pilots. Some foreign airlines that conduct parts of their ab initio training in this country do not have their pilots earn certifications and ratings here because the requirement for solo flight (which is often met by flying single-engine planes around small airports) takes time away from training that they deem more appropriate for individuals headed directly into airline cockpits.

Another example involves the emphasis on flight time in the requirements for pilot certifications and ratings. The FAA's Advanced Qualification Program for advanced training of flight crew members and other operations personnel at airlines regulated by Parts 121 and 135 emphasizes the concept of "training to proficiency." This program allows carriers to develop innovative training approaches, incorporating the most recent advances in training methods and techniques, rather than strictly adhering to traditional training programs (Steenblik, 1989:25). The committee encourages the FAA to work with industry and with collegiate aviation programs to explore whether lessons from the Advanced Qualification Program could be applied to the issue of pre-hire training, to conduct research on alternative methods of training those interested in commercial aviation careers, and to allow experimentation when appropriate.

Diversifying the Aviation Workforce

A final challenge facing aviation is diversifying its workforce. We saw in Chapter 2 that highly specialized jobs in the aviation industry are still not distributed by race and sex in proportion to the representation of minorities and women in the nation's workforce. Clearly, in a country ostensibly committed to the idea that opportunities should be open to all, the starkly low numbers of female and minority pilots and of female maintenance technicians (both minority and white) should be a serious concern. As a committee, we would be more sanguine that this concern will be addressed if we believed that economic imperatives were going to coincide with ethical ones, but unfortunately, as this chapter has shown, this may not be the case. Severe shortages that would give employers strong economic incentives to identify and hire talented people from all races and both sexes do not appear to be on the horizon. In fact, given the unique fascination of Americans with flight and the strong base provided by the recreational sector, it is possible that the competition for aviation jobs, especially pilots, will continue to be fierce. In our view, this does not diminish the moral importance of addressing underrepresentation in the aviation industry, but it does make the challenge more complex. We turn in the next chapter to an extended discussion of this issue.

5

Diversifying the Aviation Workforce

The prevalence of white men in key aviation jobs, which we describe in Chapter 2, is the legacy of both explicit discrimination in hiring and an internal culture that from the beginning of commercial aviation gave heavy emphasis to the masculine nature of flying. Aviation is not unique in this regard; its history reflects not only its own traditions but also broad societal patterns in America and the particular closeness of the industry to the military, from which many of its leaders and employees have come. Aviation is changing, as are society and the military, although the employment numbers for the industry suggest that change has come slowly. One way to ensure that aviation has the future workforce it needs is to ensure that aviation jobs are open to all members of society. There is clearly untapped potential in groups that have been historically underrepresented in the industry. Even more important is to ensure that no individual is excluded from the occupation he or she might wish to pursue on account of sex or race.

Acknowledging that formal policies once barring minorities and women from aviation jobs no longer exist, the committee also recognizes that discriminatory attitudes and practices continue in American society. We believe that no industry is exempt from their effects, and our views are shared by others who have studied these issues directly and extensively. We cite as one example the major National Research Council report on the status of black Americans, which concluded that despite "genuine progress . . . race still matters greatly in the United States. Much of the evidence reviewed in this report indicates widespread attitudes of societal racism" (Jaynes and Williams, 1989:155). The writers of that report observed that "the status of black Americans today can be characterized as a glass that is half full—if measured by progress since 1939—or as a glass that is half empty—if measured by the persisting disparities between black and white Americans since the early 1970s" (p.4). The experience of blacks in America is "unique—in its history of slavery and of extreme segregation, exclusion, and discrimination" (p. ix); it stands as the starkest reminder that the United States is nowhere near being able to declare victory in the battle to provide equal opportunities and equal treatment to all its citizens.

Aviation is clearly one of many areas in which more remains to be done. To accurately diagnose and prescribe remedies for lingering underrepresentation in this industry, it is helpful to understand something of its history.

A HISTORICAL PERSPECTIVE

For much of their existence, U.S. airlines have been peopled mostly by white males, and it is still the case that white males dominate the management and piloting ranks of the industry (Henderson, 1995:33). One might be tempted to conclude from this that minorities and women have not been especially interested in aviation, but this conclusion is contradicted by the facts.

Current scholarship on the history of aviation (summarized in Nettey, 1996, and in Smithsonian Institution books such as Brooks-Pasmany, 1991; Douglas, 1991; C. Oakes, 1991a, 1991b; Hardesty and Pisano, 1983) makes it clear that minorities—particularly blacks, on whom the most has been written—and women were just as fascinated as white men were with flying and airplanes from the earliest days. This scholarship also makes it clear that for many years roadblocks were placed in the way of blacks and women that curbed their interest or made it much harder for them to turn that interest into action.

The extent of black interest and involvement in aviation has not been fully appreciated, due in part to a lack of reporting; as Hardesty and Pisano (1983:79) note, "because of the segregated nature of black aviation during the 1920s, '30s and '40s, coverage in white periodicals is very scanty." Black pilots, both men and women, were among the pioneers of flight, despite "the widely-held notion in the aviation community that blacks lacked even the aptitude to fly" and although "blacks found themselves arbitrarily excluded from flight instruction" (Hardesty and Pisano, 1983:5). Bessie Coleman, who in 1922 became the first black pilot to earn a license in this country, had to go to France for flight training. So did Eugene Bullard, who flew with the French during World War I. Efforts began as early as the 1930s to encourage black involvement in aviation despite the barriers of segregation. Flying clubs were established for blacks (some named after Bessie Coleman, who died in an aircraft accident in 1926). One club founder also wrote a book called *Black Wings* (Powell, 1934) and urged black youth to enter aviation as a career. Blacks also established schools of aeronautics during this period, and groups such as the National Airmen's Association and the Challenger Air Pilots' Association were created to promote aviation in the black community.

Women, too, faced questions from aviation's early days—questions that are "still unresolved in many minds"—about "whether aviation was or is a 'proper' sphere" for them (Brooks-Pazmany, 1991:1). Nevertheless, women also showed early on that they wanted to be involved in the adventure of flight. Katherine Wright worked with her brothers Orville and Wilbur on their research, although she didn't fly herself. (Orville rejected all the female applicants to the flying school the brothers opened in 1910.) Some women did learn to fly before World War I; after a wartime ban on civilian flying was lifted, more of them earned their licenses. In the 1920s, women pilots barnstormed, raced, carried passengers, and set records. In 1929, women pilots decided to establish an association through which they could share their experiences, encourage other women to take up flying, and in general promote aviation. Thus the Ninety-Nines organization was born, a group that continues to this day. The first woman pilot to fly for an airline, Helen Richey, was hired by Central Airlines in 1934, but she was forced to quit after only a few months by pressure from male airline pilots. Despite this setback, women contributed to the fledgling industry by writing and giving speeches promoting air travel; the Boeing Air Transport company hired the industry's first stewardesses in 1930. During these early days, women worked as engineers, flight instructors, and entrepreneurs in various aviation-related endeavors. In 1939, the Women Flyers of America, Inc., was formed as a national flying club open to any girl or woman involved in, or even just interested in, aviation.

World War II saw the lowering of some barriers facing minorities and women in aviation, especially in the military. The Civilian Pilot Training Program and War Training Service Program provided a pool of flight instructors and a trained cadre of about 2,000 black pilots. Black military aviators who took part in the war were graduates of these programs, including the famed all-black Tuskegee Airmen (Hardesty and Pisano, 1983; Shear, 1995). Women were also trained as pilots during World War II, including about 1,000 trained as WASPs (Women's Airforce Service Pilots), who by war's end had logged nearly 300,000 flying hours and had flown nearly every aircraft in the military inventory, including bombers and fighters. Even more than blacks (whose roles were still sharply restricted by segregation), women moved into jobs throughout civilian and military aviation during this period (Douglas, 1991; Holm, 1992:64-65).

World War II military aviators, both minorities and women, faced outright discrimination in the postwar years, however, when they attempted to enter the civilian industry.

None of the black aviators who flew in the war was hired by a major airline upon return to civilian life. Blacks were hired into blue-collar service jobs as skycaps and ground handlers in the commercial aviation industry, but they were excluded from being pilots or from filling key managerial and administrative positions. It took court battles and the civil rights movement to open the cockpit to blacks in the 1960s. Continental hired Marlon Green in 1965 only after he won a case that had begun six years earlier and that had to be argued all the way to the Supreme Court (see U.S. Supreme Court, 1963). American Airlines was the first major airline to voluntarily hire a black male pilot, in 1964. The first black woman gained a cockpit job with a major airline in 1978 (Douglas, 1991; Hardesty and Pisano, 1983; Shear, 1995:A20; U.S. Congress, 1988:6).

Post-World War II demobilization also reduced the opportunities for women in the industry, as traditional attitudes about women's "proper role" reasserted themselves. No woman was hired as a pilot by a commercial air carrier until 1973. In that year, Frontier Airlines, a regional airline at the time, Eastern, and American each hired a woman pilot (Kjos, 1993:22-23; Maples, 1992:13). Until then, women were largely relegated to flying as flight attendants in the airline passenger cabin; for the most part they were also barred from mechanics positions and from certain other airline jobs (U.S. Court of Appeals, 1977:231). Minority women faced discrimination in the flight attendant ranks as well as in the cockpit. Mohawk Airlines was the first airline in the United States to break the barrier by hiring a black flight attendant in 1957 (Petzinger, 1995:38).

Opportunities gradually opened up for blacks and women, but progress was slow and often hard won. The airlines' employment practices were targeted in numerous legal challenges as well as federal enforcement actions stemming largely from the civil rights legislation and executive orders of the 1960s. The most far reaching, longest running, and highest profile court case ended only in 1995; it involved United, which was sued, along with five of the unions with which it bargained collectively, by the U.S. attorney general in 1973. The Equal Employment Opportunity Commission (EEOC) was substituted as plaintiff in 1975, and in 1976 the EEOC entered into a consent decree with the airline (U.S. District Court, 1976). The lawsuit charged United and its unions with "a pattern and practice of discrimination in hiring, termination, and other job related practices, based upon race, national origin and sex in violation of Title VII of the Civil Rights Act of 1964" (from the government's April 14, 1973, complaint as reported in U.S. District Court, 1995:2). One specific allegation accused the airline of entering into collective bargaining agreements that contained discriminatory provisions on promotion, demotion, transfer, and layoffs based on seniority. The airline and its unions agreed to undertake remedial actions and to modify their employment and membership practices for minorities and women. The decree also established incumbency employment goals, to be reached over a 5- to 6-year period, for a number of job categories, including mechanic, storekeeper and ramp service, flight attendant, customer service agent, reservation sales agent, air freight agent, management, and clerical staff; it also set hiring goals for entry-level pilot positions. Progress in implementing the goals was slower than anticipated, although incumbency goals were reached in 1989. Pilot hiring goals were the subject of further enforcement action in 1988, when the EEOC determined that some pilot trainees and probationary pilots were being subjected to race and sex discrimination (though it did not find a "pattern or practice" of such discrimination).

By 1995 the EEOC, the defendants, and the court agreed that the goals of the original decree had been met and that the decree should be dissolved. The judge noted that "[i]n short, in terms of minority and female employment, United is not the company it was eighteen years ago" and that "all of the parties, including the EEOC, now affirm that since 1988 United has been hiring minority and female pilots at ratios which consistently exceed application ratios." He went on to recognize the view of other objectors opposed to dissolving the decree, who argued that a fuller investigation would turn up evidence of continuing problems at United and that the EEOC had not adequately implemented, monitored, or enforced the provisions of the decree. The court disagreed with these allegations, though the judge also observed that the court's "conclusion is not a determination that employment discrimination may no longer be a problem at United. . . . We have no delusions that such invidious behavior has entirely ceased to plague the workplace." He ruled, however, that future remedies were properly addressed through fresh complaints to the EEOC rather than continuation of the earlier action (U.S. District Court, 1995: 8, 9-10).

Congress has also investigated the progress of airlines toward making their workforces more representative. The Government Activities and Transportation Subcommittee of the House Committee on Government Operations held hearings in 1986 and 1987 on blacks in the industry, and the full committee issued a report in 1988 (U.S. Congress, 1987a, 1987b, 1987c, 1988). Although declining to make a specific finding of discrimination against the three airlines that had been the focus of the hearings (American, TWA, and United), the committee majority stated that "it is looking for the industry to make substantial improvements in increasing representation of blacks, particularly black pilots, managers, and other professionals," and made five specific findings (U.S. Congress, 1988:23-4):

• The vast majority of black airline employees are in low-wage and unskilled positions.
• Black pilots are experiencing multiple rejections for employment from most large commercial airlines.
• For a variety of reasons, the airlines have failed to institutionalize and incorporate affirmative action into corporate policy.
• Black airline employees are disillusioned and frustrated by their companies' treatment of blacks.
• The Office of Federal Contract Compliance Programs has failed to monitor adequately the airline industry for affirmative action compliance.

The majority of members of the Committee on Government Operations recommended shifting affirmative action oversight from the Office of Federal Contract Compliance Programs to the FAA, initiating systemic investigation of at least four major carriers through the EEOC, and elevating affirmative action issues on the agendas of member airlines through the Air Transport Association, the official trade association of the airline industry.

A substantial minority of the Committee on Government Operations dissented from the committee report, citing problems with some of its findings and conclusions. (The Departments of Labor and Transportation also disagreed with the committee's recommendation to transfer affirmative action oversight of the airlines to the FAA (U.S. General Accounting Office, 1989:10), and this transfer never occurred.) Despite its dissent to the findings and conclusions, the House committee minority nevertheless agreed with much of the committee report's substance. The committee minority noted in particular (U.S. Congress, 1988:49):

> After reviewing the EEO-1 data the airlines submit as part of the equal employment opportunity requirements, rather obvious questions come to mind. If the industry is as committed to affirmative action as it claims to be, and has indeed exercised its "best efforts" to provide employment opportunities for minorities and women, why haven't their efforts yielded more significant results?

Why indeed? Our committee discussions of this difficult question led us to the conclusion that increasing the diversity of the aviation workforce, and especially broadening access to its highly skilled and most senior positions, is a task that must include but also extend beyond the industry itself. We know from available statistical data (Table 5-1) that the proportion of certificated pilots and maintenance technicians who are women is still very small; we believe that the same is true for blacks, although we could find no data to prove the point. Efforts to diversify the aviation workforce, therefore, need to begin much earlier than the point at which potential employees apply for jobs. By enlarging the pool of people interested in and qualified for aviation careers, we can potentially address two concerns simultaneously. We can increase the number of minorities and women available for employment. We can also forestall any future supply problems by ensuring that the nation's increasingly diverse workforce is being fully utilized by the aviation industry.

We therefore conclude that the challenge of improving diversity in aviation must be addressed along three dimensions. Especially in light of past efforts to discourage minorities and women from participating in aviation, efforts must be made to develop the *interest* of individuals from underrepresented groups in undertaking aviation careers. There must be equal opportunities for minorities and women to develop the *basic academic competencies* to successfully pursue aviation careers if they choose. And any remaining *barriers* must be addressed that formally or informally have a disproportionate effect on the ability of minorities or women to pursue aviation careers if they have the basic academic competencies and the interest.

TABLE 5-1   Estimated Active Airmen Certificates Held by Women, December 31, 1995 (percentages in parentheses)

	Total Certificates Held	Number Held by Women
Pilots-Total	639,184	38,032 (6.0)
Students	101,279	12,710 (12.5)
Recreational	232	16 (6.9)
Private	261,399	15,398 (5.9)
Commercial	133,980	5,694 (4.2)
Air transport	123,877	3,134 (2.5)
Flight instructor certificates	77,613	4,556 (5.9)
Mechanic[a]	405,294	3,914 (1.0)

[a]Number represents all certificates on record. No medical examination required.

SOURCE: Federal Aviation Administration unpublished data.

DEVELOPING INTEREST

Given aviation's history, it is reasonable to believe that not only must formal roadblocks be removed (as many of them have been), but also special efforts must be made to encourage blacks and women to choose a field that for so long seemed hostile to them. Individuals from underrepresented groups need to know that aviation now offers career opportunities to which they can aspire. Hoping that this message will simply filter out of its own accord flies in the face of reality, as characterized by the relative absence of role models and the public's lack of awareness about many of the contributions minorities and women have already made. As Hardesty and Pisano noted in their 1983 catalog accompanying the Smithsonian Institution's "Black Wings" exhibit on American blacks in aviation, only in the 1980s were black contributions to aeronautical history in the United States "finally being documented and recognized after years of historical neglect" (p. 3). No wonder, then, that many black youngsters and their families need help in realizing the possibilities open to them in aviation.

Fortunately, many activities aimed at involving young people and others in aviation are already under way, sponsored by the federal and state governments; by private associations like the Ninety-Nines, Women in Aviation, International, and the Organization of Black Airline Pilots; by airlines and aircraft manufacturers; and by the many professional associations that represent companies or employees involved in all aspects of commercial and general aviation. A full accounting of these programs would be impossible, but an indication of the variety of activities available is provided by *A Guide to Aviation Education Resources* (no date) published by the FAA on behalf of the National Coalition of Aviation Education. The guide describes numerous activities sponsored by the 26 members of the coalition: public information (via print, videotape, and computers), awards, scholarships, youth awareness and involvement programs, teaching kits on aviation, and others. State and federal agencies (through the National Association of State Aviation Officials (NASAO) and the FAA) have committed themselves to a formal partnership designed among other things "to further aviation education by increasing public awareness and providing programs geared to enhance all levels of our nation's education system" (National Association of State Aviation Officials, 1994:1). NASAO (1994) also reports that 41 of 44 state aviation offices responding to its survey participate in educational opportunities in some capacity, including a majority who are involved with teacher education.

It is difficult to paint a comprehensive picture of all the activities designed to spur interest in and knowledge of aviation, nor do we really know how many people are actually reached by these activities. The committee's review suggests, however, that there are several issues to which those committed to encouraging interest in aviation should attend.

Continuing Attention to Underrepresented Groups

As we have seen, blacks and women have suffered from decades of having to overcome barriers to pursuing their interest in aviation. One result is that there is less of an aviation tradition among these groups than among white men. Not surprisingly, then, voluntary programs are less apt to attract members of these underrepresented groups without special efforts to reach and recruit them. A number of aviation education programs undertake such efforts, including the FAA's Aviation Career Education Academies as well as more targeted programs sponsored by groups such as the Organization of Black Airline Pilots, the Ninety-Nines, and Women in Aviation, International.

The committee recommends that all organizations seeking to encourage interest in and knowledge of aviation focus special attention on the continuing need to reach and involve individuals from groups who have been and still are underrepresented in the industry.

In this regard, we again point out the almost total lack of information available on who is served by aviation education programs and suggest that greater efforts be undertaken to determine each program's reach and impact. This is especially important with publicly funded programs, whose administrators have a responsibility to be accountable to taxpayers for using public monies as efficiently as possible. The absence of outcome data for aviation education programs is a problem not limited to this field, but government agencies are giving the problem increased attention in other policy areas, and we recommend that aviation agencies do likewise. In the related programmatic field of aerospace, the National Research Council (1994) recently completed a study on the education programs of the National Aeronautics and Space Administration, recommending indicators and indicator systems that should be developed to gauge the outcomes of its programs aimed at elementary through graduate education. Much of the information in that report would be useful to anyone interested in developing systems for tracking the effectiveness of aviation education programs.

We also note that most aviation outreach activities focus on precollege age groups. Another opportunity for outreach and support exists in the collegiate institutions that enroll significant numbers of minority and women students in aviation education programs. As we saw in Chapter 4, a number of historically black colleges and universities, as well as several colleges with large Hispanic enrollments, offer aviation programs. These institutions, along with schools attracting women into aviation education, offer excellent opportunities for the industry to provide assistance to individuals (such as mentors and internship opportunities) to encourage their persistence in aviation and to provide support for institutions that demonstrate their ability to attract underrepresented groups into the aviation field.

The committee therefore recommends that, to increase the pool of qualified applicants from underrepresented groups for pilot, aviation maintenance technicians, and other positions in the aviation industry, airlines and other employers work aggressively to build linkages with the aviation programs at historically black colleges and universities and other schools and colleges with large minority and female enrollments.

The Importance of Partnerships

Resources of time and money are likely to be used most efficiently when the various groups involved in promoting aviation and aviation education work together. We already see many instances of this occurring, as in the National Coalition for Aviation Education; under its auspices (according to its mission statement) 26 organizations are "united to promote aviation education activities and resources; increase public understanding of the importance of aviation; and support educational initiatives at the local, state, and national levels." Another example comes from an FAA regional official, who reported to the committee that a very important benefit of the region's Aviation Career Education Academies "is the spirit of cooperation that develops between the FAA, the educational institution, the aviation industry, and the community in helping our young people succeed." This cooperation was exemplified by the number of organizations participating in the program, including the Organization of Black Airline

Pilots, the Tuskegee Airmen, the Professional Aviation Maintenance Association, the Experimental Aircraft Association, Chapter 34, the Confederate Air Force, the Air National Guard, the Aircraft Owners and Pilots Association, the Ninety-Nines, American Airlines, Southwest Airlines, Vought, Lockheed, Delta, NAS JRB Carswell, Delta Aeronautics, Interlink, the Vintage Flying Museum, several colleges and universities, Collmer Semiconductor, Inc., Lone Star Advertising, the Texas Air National Guard, Ben E. Keith Distributors, and many others.

There is more room for progress, however. A recent NASAO report (1994) indicated areas in which improvement is needed. On the basis of its survey of state aviation offices, NASAO reported that only 19 of 44 responding states had working relationships between the state aviation office and the state department of education. NASAO noted in particular the need to expand aviation education programs at the high school level, observing that most aviation education programs continue to be geared toward elementary and middle school students. (We have more to say about high school programs below.) NASAO also noted several resource opportunities that appear to be available but are not used to their fullest potential. More specifically, states indicated only limited involvement with organizations such as the Experimental Aircraft Association, Boy and Girl Scouts, 4-H, Young Astronauts, Civil Air Patrol, and the Academy of Model Aeronautics. These organizations provide valuable resources, and careful collaboration with state agencies and the federal government can advance educational opportunities through cooperative ventures (p. 27).

Using cooperative ventures to expand the reach and effectiveness of aviation education programs takes on greater importance in light of growing resource constraints in government aviation agencies. The 1994 NASAO survey found that over the three fiscal years covered, state aviation education budgets had decreased 18 percent.

Potentially more problematic, because of the agency's traditional leadership role in aviation, are cutbacks in the aviation education program at the FAA. Government-wide downsizing has led to the virtual elimination of FAA aviation education activities. We reported on the demise of the airway science program in Chapter 4. Here we focus on the FAA's other aviation education activities designed to encourage interest in aviation and aviation careers (the Aviation Career Education Academies, Aviation Education Resource Centers, teachers workshops, etc.).

The budget for these activities, which amounted to $767,000 in fiscal 1993, was reduced to $50,000 for fiscal 1996, including the value of headquarters aviation education support (these figures and the FAA explanations for the budget changes are from internal FAA documents supplied to the committee). The FAA has reduced its civilian workforce from 52,400 full-time employees at the beginning of fiscal 1993 to 47,300 in fiscal 1996. The overall discretionary budget authority for the FAA's parent agency, the U.S. Department of Transportation, is projected to decline from $39.3 billion in fiscal 1994 to $32.7 billion in fiscal 2000 (Office of Management and Budget, 1995: Table S-14). In this climate the FAA expressed the necessity "to concentrate available resources on the core mission and activities of the FAA and eliminate secondary activities, such as the aviation education program." In doing so, the FAA noted that the program "has benefits and is well received by its constituents" and expressed the hope that the program "would continue under the leadership of the states, industry, and the academic community."

Since FAA aviation education programs are often administered in partnership with state aviation offices, the committee contacted NASAO and several individual state aviation officials to learn about the expected impact of the federal cutbacks. The effects will be variable; some states plan to use their own resources to replace the assistance formerly received from the FAA, and others have had to make some painful choices to cut back or retailor their programs. Many expressed particular dismay over cutbacks affecting printing and publications, since they believe FAA publications are of high quality and have used them extensively in their programs. Formerly provided free, these publications will continue to be available in print and via computer, but FAA will charge for individual printed copies, and many state officials said they will be unable to purchase the quantities they need.

Although the committee recognizes the hardship that FAA cutbacks present, we also acknowledge that few other industries depend on government agencies for this type of assistance; other industries must build coalitions to provide these services themselves. On one hand, there are arguments in favor of continuing at least some of the FAA's aviation education activities. In particular, it seems to us that a centralized public agency is especially well suited to the task of preparing and making available high-quality and objective information on the industry and the career opportunities it offers; this kind of information is used by the many groups who share an interest in fostering aviation's growth and development. On the other hand, there are also clearly counterarguments in terms of other public priorities. The final judgment is one that the nation's elected officials must make. We recognize that the time may have come in the evolution of the air transportation industry—an evolution that has been marked by a rapidly diminishing role for government since deregulation—when at least some of the services historically provided by the FAA will have to be supported in other ways. We think that this is a development that the industry itself (perhaps through its industry representatives like the Air Transport Association and the Regional Airline Association) ought to address in cooperation with the state and private interest groups who already support so many aviation education activities.

The committee recommends that industry work in partnership with state and private groups and the FAA to maintain basic aviation education and information services. The committee further recommends that the FAA and its parent agency, the Department of Transportation, reconsider their decision to cease providing (at no cost) basic information on the aviation

industry and career opportunities that can be used by other aviation agencies and organizations to promote interest in the field.

Public Images of Aviation and Vocational Careers

The final issue we want to raise in relationship to increasing female and minority interest in aviation has to do with what we shall call image. In both small and large ways, aviation suffers from some image problems that may hamper its attempt to diversify its workforce.

A comparatively small image problem (but one not lost on women in these days of heightened awareness) is the persistent use of the term *airmen* to describe specialized aviation personnel. Official FAA publications continue to report statistics on certificates held by "U.S. civil airmen." It's time to find another term. Recently a private committee of aviation professionals was successful in working with the FAA to change the title of the FAA's *Airman's Information Manual* to *Aeronautical Information Manual* and to make the language of the document gender neutral. The first issue of the revamped publication was issued in July 1995. That committee is now working to develop proposals for making the language of the Federal Aviation Regulations likewise gender neutral.

The term *mechanic* similarly fails to convey an accurate image of the modern-day maintenance technician who often has highly specialized skills and works on aircraft loaded with extremely advanced technology. Fortunately, this terminological problem is on its way to being resolved thanks to proposed regulatory reform that will codify in the names of new certificates the occupational name *aviation maintenance technician*, which is already in widespread use in the industry. The FAA, the U.S. Department of Labor (which publishes the *Dictionary of Occupational Titles* and other occupationally specific materials), and others who provide career guidance materials should be encouraged to move quickly to adopt the new job title in their publications and presentations.

Even the job of pilot is very likely misunderstood by many. With the advent of glass cockpits and fly-by-wire technology, flying the most advanced aircraft made is less a physical challenge than an exercise in systems management and crew leadership. Those unfamiliar with the modern airplane or steeped in the movie imagery of the "Top Gun" military fighter pilot may harbor outdated perceptions of the pilot's job and accompanying beliefs that women in particular are less well suited to performing it.

Another of aviation's image problems is the narrow view most people have about who works in aviation (Eiff et al., 1993). From the earliest days of aviation, the public's imagination was fired by the daring exploits of pilots, and, even in the present, pilots continue to be the first group most people think of when they think about jobs in aviation. Taking a different approach to the aviation workforce by focusing on those occupations that require elaborate formal training and licensing to ensure the public safety, one again thinks of pilots, along with aviation maintenance technicians.

As we have emphasized in this report, however, these specialized aviation occupations represent only a small portion of the aviation workforce and account for quite a small number of workers (roughly a quarter of a million people). Moreover, pilot jobs in particular have been and may well continue to be highly sought after; it is unclear that there will ever be the kind of shortages that so worried policy makers in the late 1980s. Some public officials recognized the dangers of an overly narrow focus in their aviation education efforts. An aviation/aerospace task force report to the Oklahoma State Regents for Higher Education (1994) stated that respondents to an industrial survey warned the task force not to fall into "the pilot-shortage trap, deeming other career areas to be more critical." The task force found that Oklahoma was producing an overabundance of entry-level mechanics (with airframe and power plant certificates) and primary commercial pilots and urged the state to spend less on this type of training and more on training for other aviation and aerospace fields.

Individuals interested in aviation are very likely to find it much easier to enter the industry if they take and are encouraged to take a much broader view of the opportunities available. As the Oklahoma report suggests, aviation and aerospace are closely linked and together offer employment in many areas requiring advanced knowledge and skills. Even within aviation and its affiliated services (like airport operations, air traffic control, and safety inspection) there are good jobs that may go unrecognized. The Aircraft Owners and Pilots Association is one group that has shown how students can be exposed to the wider picture. The group publishes a "help wanted" brochure on careers in aviation that shows career titles, salary ranges, educational requirements, and typical employers for 62 career fields in 10 major categories; the 10 are: pilot careers, airline and airport operations, airline and airport services, aircraft and systems maintenance, aircraft manufacturing occupations, scientific and technical services, law-related services, health services, office professionals, and food services.

A much deeper and more difficult image problem for aviation is the low regard with which technical jobs and vocationally oriented education and training are often held in this country. Several prominent reports (Marshall and Tucker, 1992; National Center on Education and the Economy, 1990; William T. Grant Foundation Commission, 1988) have highlighted America's comparative inattention to how it trains its workforce and especially to the unfocused, outdated, and generally second-best education frequently provided to high school students not studying on the academic track. These reports have struck a nerve in large part because of widespread fears that the United States is losing its competitive advantage in the global marketplace. The 1990s have seen significant reform of vocationally oriented education programs at the federal, state, and local levels and the creation in many states of comprehensive workforce development agencies and boards. Most of these reforms are new, and their outcomes are as yet uncertain.

Reform efforts aim in part to prove that technical and vocational education can be as rigorous as traditional academic coursework and, in doing so, to raise the status and appeal of preparation for technical occupations. The hope is also to provide educational alternatives that will appeal to students who in the past found school irrelevant or uninteresting and dropped out before gaining the skills that will be needed in a fast-paced, competitive, technologically driven workplace. Many of them are black and Hispanic students who come from economically and socially disadvantaged backgrounds, whose high school completion and college enrollment rates have historically trailed those of whites.

Aviation, with its inherent appeal, clearly has much to offer to these reform efforts. Numerous programs demonstrate how aviation-oriented education can be used to encourage interest in mathematics and science, to spur students to stay in school, and to improve coordination between high school education and technical training in community colleges, as well as to boost interest in aviation careers.

Survey results published in 1994 (Mitchell) identified 34 secondary magnet schools offering some kind of aviation or aerospace technology program. Though the first whole-school aviation magnet program dates back to 1936 (Aviation High School, Long Island, New York), the concept appears to be growing and spreading. Many of the schools reported that they "assertively" recruit minorities.

Opportunity Skyway, a national career preparation program headquartered at a historic airport site in Prince Georges County, Maryland, uses a coalition-building approach to developing high school, and to a lesser degree middle school, programs. By relating school curriculum to a broad range of careers in aviation, this program aims to keep students in school, maintain their interest in learning, and focus them on careers. A nonprofit organization, Opportunity Skyway was founded in 1990 by the Prince Georges Private Industry Council and today has programs operating or under development in 13 states. Through the program, students have the opportunity to learn aeronautics integrated with math, science, and other subjects, to attend private pilot ground school and receive flight training, and to participate in internships and apprenticeship-type workplace opportunities.

National school reform efforts have featured new school-to-work approaches, such as youth apprenticeships to link schools and employers and technical-preparation programs aimed at coordinating technical-preparation curricula in high schools and community colleges. Although we are not aware of any strictly aviation programs that embody these approaches, we do know of an example in the related field of aerospace that shows what can be done. The Middle Georgia Aerospace Youth Apprenticeship Consortium prepares high school and post-secondary students for work through a combination of applied academics and structured workplace learning. McDonnell-Douglas and Northrop-Grumman cooperate with three county school systems and three technical institutes. Students are selected to participate in the program after their sophomore year in high school and follow a tech-prep curriculum designed for aircraft manufacturing technology. The aerospace companies provide a paid 2-week internship at the end of the apprentice's junior year and a more intense, paid 8-week internship at the end of the senior year. The program leads to certification as an aircraft structural specialist, which is a nationally recognized certification, upon completion of the apprenticeship program at a technical institute. Students have the option of pursuing an associate's degree. While no job is guaranteed, the successful apprentice receives priority in consideration by participating employers. Of 71 apprentices currently in the program, 32 are white males, 12 are white females, 15 are minority males, and 12 are minority females (information supplied to the committee by Alicia Long of the Middle Georgia Aerospace Youth Apprenticeship Program).

In aviation, because many entry-level jobs are with fixed-base operators and other relatively small employers, school-to-work programs may require special leadership by school and industry groups and may need adaptation to the special circumstances of these employers. The committee urges groups in areas in which entry-level employment jobs are numerous to explore the advantages that such programs might offer.

We also urge local, state, and federal authorities and others involved in promoting these programs to give special attention to follow-up and evaluation. Too often in the course of researching this report, we have identified promising initiatives of one sort or another (school-industry collaborations in Chapter 4, programs to spur interest in aviation in this chapter), but we really can't say whether the programs make any difference or not because only informal and haphazard efforts are made to evaluate program outcomes. Improving the diversity of the aviation workforce, along with the other goals of the programs we have been describing, has proven hard to accomplish. The chances of succeeding can be increased by taking seriously the need to determine whether particular efforts do or do not help change people's behavior.

The committee recommends that the responsible agencies and groups work to create more accurate public understanding of modern aviation careers and acceptance of the technical education needed to prepare for them.

BASIC ACADEMIC COMPETENCIES[1]

In broadening the diversity of its workforce, aviation shares a dilemma with other industries dependent on advanced and rapidly changing technologies: un-

---

[1] In preparing this section, the committee has drawn in part on the work of the Committee on the Feasibility of a National Scholars Program of the National Research Council's Office of Scientific and Engineering Personnel.

equal preparation of minorities and women in mathematics, science, and technology-related subjects. For all these industries, part of the answer to diversifying their workforces is increasing the pool of individuals from historically underrepresented groups who have the basic academic competencies that will allow them to pursue more specialized education and training.

The problem of poor preparation in math and science and its implications for employment in scientific, mathematical, engineering, and technical jobs has been widely studied. Although it is not clear how many of these research findings are applicable to aviation careers, what is clear is that a solid grounding in precollege mathematics and science is essential for individuals wishing to pursue many aviation-related training opportunities. Committee members who work in aviation education note that students with poor precollegiate training in math and science will not be adequately prepared to enroll in many collegiate aviation programs. Engineering (including aerospace engineering) will also be closed off to students without adequate backgrounds in math and science. Furthermore, one airline that has publicly described the criteria it uses to select pilot trainees has indicated that proficiency in scientific subjects at school is a good predictor of success in flight training (Acton, 1989:74). The rapid advance of technology in aviation requires not only that students have the basic mathematical and scientific competencies to undertake initial training but also that they be able to continue learning throughout their working lives.

College graduation statistics show that minorities (especially blacks, Hispanics, and American Indians/Alaskan natives) and women major in science, mathematics, engineering, and technology fields at comparatively low rates (Table 5-2), but the problem of underrepresentation in these fields starts much earlier. The inclination toward a career requiring math and science abilities has often been determined by the time an individual finishes high school (J. Oakes, 1990a; Transportation Research Board, 1992). Therefore the opportunity to develop these abilities in school (kindergarten through 12th grade or K-12) becomes of paramount importance in understanding the readiness of underrepresented groups.

Research has provided a great deal of evidence about differences in mathematics and science participation and achievement among minorities and women at the K-12 level (National Science Board, 1996:1-2). The good news is that attention to these issues appears to be paying off: achievement is improving and gaps are narrowing. Sex differences in mathematics achievement test scores at the precollegiate level have all but disappeared, and sex differences in average science achievement test scores have diminished, though boys continue to score higher at each grade level tested. Standardized mathematics and science test scores of black and Hispanic youth continue to lag those of non-Hispanic white youth, but the differences decreased from the late 1970s to 1992. Moreover, "since the average scores of all groups have increased over this period, the convergence is real improvement and not a 'leveling down' of scores toward greater equality."

TABLE 5-2 Baccalaurate Degrees Conferred by Institutions of Higher Education by Racial/Ethnic Group and Sex of Students for Selected Fields of Studies, 1992-1993

Major Fields of Study

	Computer/ Information Science	Engineering	Engineering- Related Technologies	Mathematics	Physical Sciences and Science Technologies	Transportation and Material Moving	All Fields
Total degrees conferred	24,200	61,973	15,904	14,812	17,545	3,930	1,159,931
Percentage awarded-total	100.0	100.0	100.0	100.0	100.0	100.0	100.0
White non-Hispanic	68.2	73.9	81.3	80.1	81.4	88.6	81.7
Black non-Hispanic	9.3	4.2	6.7	6.6	4.8	4.1	6.7
Hispanic	3.6	3.7	3.9	2.9	2.5	3.5	3.9
Asian/Pacific Islander	9.5	10.6	4.8	6.3	6.4	1.9	4.4
Other	9.4	7.6	3.2	4.1	4.9	2.0	3.3
Percentage awarded-men	71.9	84.2	91.1	52.8	67.4	89.3	45.7
White non-Hispanic	52.0	63.0	74.7	42.2	56.2	79.4	37.5
Black non-Hispanic	4.4	2.9	5.5	3.0	2.2	3.3	2.5
Hispanic	2.3	3.0	3.5	1.7	1.7	3.1	1.7
Asian/Pacific Islander	6.3	8.5	4.4	3.4	4.0	1.7	2.2
Other	6.8	6.7	3.0	2.6	3.3	1.8	1.8
Percentage awarded-women	28.1	15.8	8.9	47.2	32.6	10.7	54.3
White non-Hispanic	16.2	10.9	6.6	38.0	25.1	9.2	44.2
Black non-Hispanic	4.9	1.3	1.2	3.6	2.7	0.8	4.2
Hispanic	1.2	0.7	0.4	1.2	0.8	0.4	2.2
Asian/Pacific Islander	3.2	2.1	0.5	2.8	2.4	0.2	2.3
Other	2.6	0.8	0.3	1.6	1.6	0.2	1.4

SOURCE: 1995 Digest of Education Statistics, Table 257 (available on the Internet).

Clearly, though, there is still work to be done. A brief review of what has been learned so far indicates the challenges that remain. For minorities who are also at a socioeconomic disadvantage (especially blacks and Hispanics, on whom the most information has been collected), disparities in educational resources and opportunities to learn are apparent as early as elementary school and persist throughout the K-12 years. For girls, resources and opportunities are less the issue than are social and cultural mores that by high school have left many of them disaffected from mathematics and science.

Minorities

Math and Science Exposure in Elementary School

Research has shown that minorities have fewer opportunities to gain exposure to math and science skills. Reasons include interest, time devoted to math and science instruction in predominantly minority schools, teacher treatment or encouragement, and the impact of programs designed to develop interests and competencies.

Differences in attainment are evident by the third grade in mathematics, and they grow thereafter. Nationwide, only a small percentage of elementary schools have substantial science and mathematics resources. Moreover, minority students have less access to computers in elementary school (J. Oakes, 1990b; National Science Board, 1996). Minority and lower income students also lag behind wealthier students in their access to the Internet. A recent survey of 917 public schools found that only 31 percent of low-income schools (defined as those in which 71 percent or more of the students were eligible for free or reduced-price school lunches) had Internet access, compared with 62 percent of schools in which 11 percent fewer students were eligible for subsidized lunches (National Center for Education Statistics, 1996).

Interestingly, however, schools with the largest percentages of black and Hispanic students and those in inner cities are reported to devote the most time to mathematics; this additional time spent on mathematics has not been gained at the expense of science instruction because the schools involved had longer class days (J. Oakes, 1990b). Oakes notes that most of these schools are eligible for federal Chapter 1[2] funds targeted at improving disadvantaged students' basic skills. She found, however, that federal assistance did not seem to directly affect the time spent on mathematics instruction, but it did affect the time teachers said they spent on science. Oakes speculates that the extra resources provided by Chapter 1 funds—specialist teachers and instructional materials—may enable

---

[2]Now called Title I. The reference is to the first part of the Elementary and Secondary Education Act, first passed by Congress in 1965 and amended periodically since then.

schools to free up more teacher time and resources for science instruction. She speculated that "raising students' mathematics achievement is thought to be important enough to command a large share of time *with or without Chapter 1 funding* "(p. 29, emphasis in original). In contrast to these school differences, she did not find differences in the amount of time students in low-, average-, and high-ability classes spent on science and math lessons. Because schools often regulate the minimum amount of class time for various subjects, teachers may have little discretion over the amount of time they spend with students of different ability levels.

According to J. Oakes (1990b), the time teachers allocate to lessons is far less important than the time students are actually engaged in tasks. Achievement data showing a slow but steady decrease in the black-white mathematics gap in the elementary grades over the past decade suggest that some factor in basic skills instruction may be having a modest benefit.

Math and Science Exposure in Middle and Junior High Schools

The middle school years usually determine whether a student will participate in the academic track, traditionally a prerequisite for access to advanced math and science courses. There appears to be no significant racial or ethnic difference in eighth-grade students' attitudes toward math and science. In fact, a greater percentage of black, Hispanic, and American Indian eighth graders than of white said that they looked forward to their math and science classes. At the middle/ junior high school stage, however, minority achievement begins to decline. These years are a time of great developmental change for youngsters (Clewell et al., 1992); they also appear to be a time when obvious educational inequities exist between minority and majority students.

J. Oakes (1990b) has documented numerous differences between schools with student populations that are more than 90 percent white and schools with similar populations of minorities. Science programs at the largely white schools were more extensive (nearly 4 classes per 100 students) than at the largely minority schools. No significant race-related differences were found in the size of mathematics programs. Many junior high and middle school students are required to take only one semester of science; additional science courses are either electives or are recommended for high-achieving students. Oakes concluded that high socioeconomic status (SES) schools and predominantly white schools have either greater science requirements or offer greater numbers of optional science courses.

The lower-SES schools in the Oakes study had significantly larger class sizes in mathematics than did the most affluent schools (averaging 26 and 23 students per class, respectively). Junior high schools serving the largest concentration of low-income students allocated less time and fewer teachers to mathematics. Oakes concluded that the advantages of more time on mathematics instruction in elementary schools serving large numbers of low-income and minority students may disappear when students reach middle school or junior high. Differences in teacher attitudes are also evident at this stage (National Science Foundation, 1994). Teachers at economically advantaged schools typically report high morale and positive attitudes about students, while those at economically disadvantaged schools report substantial difficulty motivating students. Only 7 percent of white eighth graders attended disadvantaged schools, compared with 15 percent, 36 percent, 39 percent, and 40 percent of their Asian, black, Hispanic, and American Indian peers, respectively.

Almost 46 percent of white eighth graders were taught math by teachers who had majored in math, compared with 44 percent, 40 percent, 33 percent, and 30 percent of Asian, black, Hispanic, and American Indian students, respectively. For science, 53 percent of Asian eighth grade students had teachers who majored in science, compared with 49 percent of whites and blacks, 47 percent of Hispanics, and 40 percent of American Indians (National Science Foundation, 1994).

Math and Science Exposure in High School

Minority student interest in math and science continues to compare favorably with that of other students into the high school years. White 12th graders were less likely than their black, Hispanic, and American Indian counterparts to say they liked math, and less likely than their Hispanic and American Indian peers to say they liked science (National Science Foundation, 1994). Nevertheless, disparities in resources available to and achievement of minority students are again apparent.

In contrast to the similarities in teacher qualifications at the elementary school level, teacher qualifications in secondary schools differ substantially across school types. Teachers at schools with predominantly economically advantaged white students and teachers at suburban schools tend to be the most qualified (J. Oakes, 1990b).

This study also reveals that the principals of low-SES high schools were less likely than those of higher-SES schools to report having had computers available for instructional use (77 versus 95 percent, respectively). Moreover, teachers at low-SES and inner-city schools reported that computers were less readily available at their schools, or, if they were available, they were difficult to secure for use in instruction. More recent evidence (National Science Foundation, 1996) confirms that science teachers in high schools with high minority enrollments continue to be more likely to report needing computers, as well as other equipment, that is unavailable. There are also some recent findings, however, that black students are becoming more likely than white students to use computers in science and mathematics classes. In addition, in 1992 a higher percentage of black than white students reported having ever been taught a computer skill or a programming course.

While considerable evidence attests to the importance of high school course taking for mathematics and science achievement, black, Hispanic, and American Indian children continue to complete fewer advanced science and mathematics courses by the time they graduate from high schools than do Asian American or white students (Peng et al., 1994; National Science Board, 1996). Secondary schools vary considerably in the time they allocate to science and math. Differences in course-taking patterns are related to school characteristics: location, type and size, and socioeconomic status and ethnic status of the student body. Course taking is generally lowest for extremely rural schools, followed by disadvantaged urban schools. The pattern of lower-level course taking for students in schools with more low-income families (defined as a higher percentage eligible for subsidized lunches) was consistent across all courses. Lower percentages of students took courses as the percentage of non-Asian minority students increased. Except for chemistry and biology, schools with the highest proportions of minority students showed a slight advantage over schools with minority enrollments between 50 and 74 percent (National Science Foundation, 1993).

Students who take more advanced math and science courses in high school have been shown to receive higher math scores on college entrance exams (National Science Foundation, 1994) Participation in college preparatory mathematics (geometry in particular) is strongly associated with college attendance (Pelavin and Kane, 1990).

Data from the 1992 National Assessment of Educational Progress indicated that about 73 percent of white 17-year-olds had taken geometry or higher-level mathematics courses, compared to about 60 percent of black and Hispanic students. This reflects a doubling in the percentages of black and Hispanic students taking these courses since 1982. However, the math scores for whites in 1992 were still significantly higher—by over 20 points—than for both blacks and Hispanics (National Science Board, 1996). These differences in course taking and in the scores of 17-year-olds may be traceable in part to developments in the middle school years, when higher-scoring students are placed in higher-level mathematics classes and minority students are more likely to be placed in lower-level and remedial courses (J. Oakes, 1990a).

Teacher Attitudes Toward Minority Achievement

Several observational studies of classroom interaction have shown that teachers have different expectations for minorities and treat them differently. Trujillo (1986) found that majority teachers, in interaction with black and Latino students, spent less time responding to their questions, asked them less complex questions, and did not give them as many hints and clues for improving performance as they did for majority students. Trujillo suggests that this pattern is due to low expectations. Teachers reserve their best attention for students who are perceived to be the best and the brightest and give other students short shrift. However, earlier work by Rubovits and Maehr (1977) suggests that teachers may also treat bright students differently depending on race. They found that, although bright white students received the most attention from their teachers, bright black students received the least attention from their teachers. Furthermore, the higher a teacher scores on a measure of prejudice, the more likely he or she is to ignore bright black students.

Studies by Contrearas and Lee (1990) and Kahle and Lakes (1983) suggest that differential expectations of student performance based on race extend into science and math classrooms. A further demonstration that teachers do not have the same expectations for all students comes from the Equity 2000 program of the College Board (College Board, 1995a). Based on the findings of Pelavin and Kane (1990) and others, this project attempts to reduce ethnic differences in academic achievement and college attendance by increasing participation in algebra and geometry by all students (eventually reaching 100 percent) in 14 sites around the country. Project sponsors have been surprised by the unanticipated attitude they encountered among teachers and others that all students were not capable of mastering these courses. Although the Equity 2000 findings do not indicate that expectations about who can and cannot achieve in science and math are based on the race of the students, they are disturbing to many who believe that low expectations in fact often hold minority students back.

Women

For women, particularly nonminority women, broad social and cultural mores, stereotypical attitudes, and societal biases are paramount in discouraging and deterring them from studying science and mathematics (J. Oakes, 1990a; Vetter, 1994). These issues are more important for women than are the long-standing economic and educational inequities that confront underparticipating minority populations, because student socioeconomic status is not associated with sex. Although the kind and quality of the classroom experiences of girls may differ from those of boys, the elementary and secondary schools they attend are not inferior to those of boys; girls, in theory, have access to the same curriculum as boys.

The attitudes of girls and boys diverge early, however, particularly in the area of science. Even in elementary school, girls show less positive attitudes toward science and science careers than do boys and report fewer science experiences. These patterns are also found in middle and high schools. Moreover, by high school, researchers found girls expressing more negative attitudes about mathematics as well (J. Oakes, 1990a), although this may be changing, with girls now virtually achieving parity with boys in mathematics course taking in high school.

Whereas girls' mathematics achievement as measured by the National Assessment of Educational Progress in 1992 was virtually the same as boys' for 9-

year-olds and 13-year-olds, and only slightly lower for 17-year-olds, girls continue to lag behind boys in science achievement at all ages. This parallels course-taking patterns, with virtually no difference in mathematics, but a sizable gap in science, in which girls are slightly more likely to take biology and chemistry but noticeably less likely to take physics (National Science Board, 1996).

J. Oakes (1990a) reports numerous studies suggesting that boys see mathematics as more useful than girls do and that substantial numbers of students think that studying science is more important for boys and that boys understand science better. Although some of these studies are 10 to 20 years old, a study conducted in 1989-1990 showed girls more likely than boys to say that they failed to take a mathematics or science course in their senior year because they didn't like the subject and/or because they had been advised not to (National Science Foundation, 1994). Research using the same database indicated that male 10th graders were more likely than females to report talking to their parents about science and technology issues. In addition to expressing less liking for math and science and less tendency to talk about them with parents, girls are less confident about their ability to succeed in mathematics and science, even when they are in fact equally able. They are also more likely to perceive role conflicts if they aspire to mathematics and science careers (J. Oakes, 1990a). Vetter (1994) reports that girls who are mathematically gifted are less likely than similarly qualified boys to be selected by teachers and principals for special mathematics programs and that those who are selected and who enter are less likely to be encouraged by parents, peers, and teachers to continue in such programs.

Finally, there are sex-related differences in classroom experiences. J. Oakes (1990a) summarizes studies showing that boys tend to have more experiences with science equipment and different kinds of science instruments in elementary school classrooms, that elementary school teachers interact with boys more frequently than with girls during instruction and provide greater encouragement for boys in both science and mathematics, and that high school teachers were observed initiating more interactions with boys and providing more specific feedback to them. A National Science Foundation report (1994) similarly reports on research showing that math and science teachers make more eye contact with boys than with girls in their classrooms, pay more attention to boys, respond to wrong answers given by boys by challenging them to find the correct answers, and give sympathy to girls. Boys also tend to get to operate the equipment and perform science experiments, whereas girls tend to record the data and write up the results.

Improving Mathematics and Science Competencies Among Minorities and Women

The task of improving mathematics and science competencies so that minorities and women are prepared for aviation careers if they so choose is bound up in much larger questions of reforming K-12 education and of improving the preparation of all students for science, math, engineering, and technology careers. The remedies lie in efforts to improve these larger systems. Some important things have been learned from past efforts, but there is also a widespread belief that most initiatives to date have had modest if any success.

Over the past quarter of a century, numerous programs have attempted to increase the participation of minorities and women in scientific, mathematic, engineering, and technical careers. Many of these programs emerged from the civil rights movements of the 1960s and 1970s and were aimed at reducing the extreme underrepresentation of women and minorities in these fields. At first, the programs were mainly local initiatives based on locally identified needs. Once these programs had been in existence for some years, however, they began to attract national attention and federal and foundation funding (Clewell et al., 1992).

K-12 intervention programs tended to utilize approaches different from those of the traditional educational system, because of the recognition that formal education had failed to address the problem in the past. Consequently, although these intervention programs may be in-school programs offered during the school day, they operate separately from the school system and have a number of distinguishing characteristics (Clewell et al., 1992).

- They focus on counteracting some educational inequity experienced by one or more of these groups.
- They use innovative instructional techniques, materials, and curricula.
- They often address one problem area, such as math or science, rather than the whole range of educational problems.
- They engage in activities that address many aspects of the problem, not just those that are focused on achievement.
- They employ multiple strategies to obtain their objectives.
- They are sensitive to the needs of the groups that they intend to serve and develop their intervention approaches around those needs.

Although the first intervention programs targeted students in high school and college, there has been a growing awareness that the factors impeding minority access to scientific, mathematical, engineering, and technical careers are present long before high school and that intervention to increase the talent pool is best undertaken before the ninth grade, whereas strategies to decrease attrition from the pool should be targeted at all points in the process.

These lessons are important in guiding the evolution of programs to improve the academic preparation of minorities and women in mathematics and science, but educators agree that there are no quick fixes for the problem of underrepresentation in scientific, mathematical, engineering, and technical careers (Matyas and Malcom, 1991; Institute of Medicine, 1994; Malcom and Chubin, 1994) or for the larger problems of K-12 education. The long-term goal must be the fundamental reform of the education and training system that presently relegates so many students to mediocrity or failure. Indeed, this is the goal of many ongoing efforts: by the federal government (the Goals 2000, School-to-Work, National Science Foundation Systemic Reform initiatives); by state and local governments (standards-based reform); and by private groups and coalitions (Equity 2000 of the College Board, the National Action Council for Minorities in Engineering).

Whereas the ultimate goal of educational reform should obviate the need for special programs targeted on minorities and women, such targeted programs continue to serve necessary purposes. There is a growing consensus, however, about the need for improved collaboration (Institute of Medicine, 1994; Board on Engineering Education, 1994; Quality Education for Minorities Network, 1990). Efforts to improve collaboration are occurring. A significant example at the federal level grows out of the commitment of the National Science and Technology Council to produce a human resources development policy and the declaration that all federal agency educational programs in science, mathematics, and engineering will have as one measure of their success their impact on increased participation by underparticipating groups (Office of Science and Technology Policy, 1994). The strategic plan prepared by the NSTC Committee on Education and Training states that, beginning in fiscal 1996, the committee "will have in place and implement a coordinated strategy designed to foster increased participation of underrepresented groups" (National Science and Technology Council, 1995).

It is the conviction of this committee that all those interested in improving access to and the preparation of young people for aviation careers (including federal and state agencies, schools and colleges, the industry, and aviation interest groups) should recognize the linkages between this goal and the larger goals of improving K-12 education and increasing minority and female representation in scientific, mathematical, engineering, and technical careers. Especially at the precollegiate level, the purpose of motivating and preparing students to pursue further studies that require a solid mathematics and science foundation is widely shared. Thus aviation's interests will be well served by effective efforts to reform the general preparation of students in math and science, as well as to improve academic performance more broadly. Where specialized aviation programs exist at the K-12 level, they are likely to be most effective when they collaborate with rather than operate independently of systemic efforts to improve educational performance.

The committee recommends support for efforts to improve the general preparation of elementary and secondary school students in mathematics and science and stresses the continuing need to focus special attention on improving opportunities for and the academic achievement of minorities and women. The committee also recommends that those responsible for

specialized aviation programs at the K-12 level collaborate with larger systemic efforts to improve educational performance.

BARRIERS

Formal barriers restricting the employment of blacks and women in aviation may be gone, but there are still obstacles to be overcome. One clear-cut obstacle is the cost of training. More subjectively, we, like the judge quoted earlier in the United consent decree case, are under no delusions that invidious behavior has entirely ceased to plague the workplace. The more intangible barriers that result from such behavior, such as those that may affect how individuals are selected for jobs or what kind of climate they encounter in schools and businesses, also need continuing attention.

Costs

The costs of specialized training for aviation careers, particularly for pilots, are substantial. High costs are likely to pose a special barrier for students who come from households with below-average incomes. Many black and Hispanic households have incomes substantially below those of whites and Asian-Americans (Table 5-3).

Flight training can add as much as $7,500 annually to the regular cost of postsecondary education, depending on the college attended. Regular tuition, room and board, and other expenses for academic year 1994-1995 at U.S. colleges averaged $6,454 at public four-year institutions and $15,528 at private four-year schools[3] (College Board, 1995b:6).

When flight training is part of a student's regular collegiate program of study, it is considered a "lab fee" and can be included along with tuition and fees in determining eligibility for publicly funded student assistance that is awarded on the basis of the applicant's financial need. The federal government provided approximately $35 billion annually in aid to postsecondary students, primarily through Pell Grants ($6 billion) and student loans ($26 billion). These programs were established to equalize access to higher education in this country (as were state student aid programs that exist in every state and amount to $3 billion annually). Aviation students along with their colleagues in other fields of study benefit from the opportunities opened up by this assistance. At the undergraduate level, public programs do not provide aid based on the course of study undertaken.

Although these programs are helpful, they do not remove all the cost barriers

---

[3]Annual costs of attendance at public and private universities that offer graduate programs are higher than at colleges offering undergraduate programs only: $7,035 and $21,152, respectively, in 1994-1995.

TABLE 5-3   Household and Per Capita Incomes by
Racial/Ethnic Status, 1994

	Median Income (1994 dollars)
Household	
White non-Hispanic	$35,048
Black	20,032
Asian/Pacific Islander	39,329
Hispanic	23,472
Per capita	
White (includes Hispanic)	17,230
Black	10,116
Asian/Pacific Islander	16,093
Hispanic	9,056

SOURCE: Bureau of the Census, data compiled from information collected in the March 1995 Current Population Survey (available on the Internet).

facing students from economically disadvantaged backgrounds. The maximum annual Pell Grant was $2,470 in academic year 1996-1997; the maximum annual student loan for an undergraduate still financially dependent on his or her parents is $5,500.[4] For students whose parents can provide little assistance, clearly additional financial aid funds will be needed. Colleges provide some funds from their own resources; industry and aviation interest groups also provide scholarships. But we have been told that the costs of education still pose a barrier for many students and discourage some from undertaking expensive options such as flight training. We have also been told that privately sponsored scholarship programs are sometimes unwilling to allow the incremental costs of flight training to be included in the costs of education they will cover.

The situation of students in flight programs who must borrow relatively large amounts of money and then face years in low-paying jobs while they build up flight time and experience poses another question: Can they repay their loans? Recent changes in loan programs are likely to be helpful. Traditionally students faced loan repayment schedules that required them to begin payments 6 to 9 months after leaving school and to make equal monthly payments over a 10-year period. This schedule was not well suited to the earnings patterns of borrowers with very low earnings in their first years in the labor market, even if their later earnings prospects were good. Program changes made in 1994 now permit more flexible repayment options, one allowing graduated payments that increase in

---

[4]Actually, this annual limit applies only to juniors and seniors; freshmen and sophomores are limited to $2,625 and $3,500, respectively, annually under the largest federal loan program.

amount every 2 years, presumably as income grows. An income-contingent repayment option sets annual repayment amounts based on the borrower's adjusted gross income and the total amount of debt. Student borrowers now also have the option to extend their loan repayments for up to as much as 30 years.

These new loan options should ease the way for students who have to borrow to pay for their undergraduate education. Nevertheless, because flight training is expensive, there is no real low-cost college alternative available to would-be pilots, as there is for students in other programs who might choose to enroll in low-tuition public community and four-year colleges. This fact suggests that financial barriers still exist, especially for students from lower-income families who are generally thought to be more reluctant to borrow than students from families with average or above-average incomes (U.S. General Accounting Office, 1995). In addition, Embry-Riddle Aeronautical University, with the largest undergraduate population of future airline pilots in the country, reports (in a communication to the committee) on the basis of a preliminary survey of recent graduates that their black and other minority pilots appear to be having more difficulty finding employment after graduation than their white male and female counterparts. This is a disturbing finding indeed, not only for the ability of these students to repay their loans but also for their prospects of moving up through the post-college steps that lead to jobs in the aviation industry.

Serious financial obstacles may exist in any event at this transition phase between initial flight training and employment by an airline. Student loans are not available to finance the many hours of flight time that pilots must accumulate after college to be ready to apply for an airline job. These jobs generally require at least 1,500 hours of flight time and a minimum of 250 hours of multiengine time. Pilots who must pay to fly multiengine planes can face per hour flight costs between $150 and $300. We also saw in Chapter 4 that a tight labor market for pilots has allowed some airlines to require that pilot candidates themselves pay additional training expenses that once would have been paid by the company, such as training for type ratings and post-hire training on company planes and procedures. Such training can cost $10,000 or more.

The committee recommends the establishment of financial assistance programs to help applicants for pilot positions meet the costs of flight and transitional training. We think the aviation industry should take the lead. If employers actively participate now in efforts to help individuals who want to fly obtain the necessary training, they can help themselves avoid the far larger costs they would incur if they had to provide training themselves (through something like ab initio programs). Employers might consider making grants or loans to individuals in training programs or donating funds for financial aid to collegiate aviation programs. Industry trade associations might develop such programs on behalf of airlines in general, perhaps pooling resources from their members. Trade associations might also be appropriate sponsors of loan programs for pilots trying to meet the costs of transitional training, to provide access to loans for individuals whose flight schools have not developed their own lending arrangements. Airlines that require applicants to pay for their own airline-specific initial training should consider making loans available to cover the costs of this training.

Selection Procedures

Investigating barriers that may restrict access to aviation jobs for minorities and women naturally raises the question of selection procedures. The issue arises especially for professional pilots, who are still predominantly white and male. There are actually two questions: What procedures do the airlines use to choose pilots? Are there elements of the pilot selection process that create obstacles for nontraditional candidates?

We were not able to investigate these questions in depth, but in some initial explorations we were surprised to discover how little information is available in the public domain about how pilots are selected for civilian employment. In its pilot's guide, FAPA (1993) gives some illustrative examples of instruments used by airlines. FAPA (until business ceased in late 1996) and similar services help individual job applicants, for a fee, to learn about the selection procedures of specific airlines based on the experiences of previous clients who have been through the process. Books (often written by pilots—e.g., Griffin, 1990; Mark, 1994) describe the process as the pilots have experienced it both as job candidates and as airline employees serving on selection panels. But the information appears to us highly unsystematic and the process still relatively mysterious.

A more transparent process would seem helpful both for training schools and for would-be pilots. It could also help individuals judge when they are ready to apply for airline jobs. This is important, since we have been told that airlines generally do not call people back for a subsequent interview if they have once been rejected. Thus, applying too soon relative to the carrier's criteria can result in a permanent rejection from that company.

The committee recommends that airlines formalize and publicize their hiring criteria so that schools can develop appropriate programs of study and individuals can make informed decisions about training and career paths.

It appears that pilot selection criteria are also in need of review and that changes suggested by new job task analyses and research on crew performance might improve selection procedures and also reduce barriers for women and minorities. Addressing this issue directly is complicated by the fact that not only details of airline selection systems but also the underlying research basis is considered proprietary by the airlines. Very little information is available in the published research literature on civilian pilot selection (Damos, 1995; Hunter and Burke, 1995). Virtually all of the available literature focuses on military pilot selection and training. To the extent that civilian airlines prefer military-trained pilots when they can get them, however, military selection procedures are of

TABLE 5-4   Air Force ROTC Pilot Training Candidate Pool and Selection
Rates by Sex and Ethnicity, 1981-1992 (percentages in parentheses)

	Air Force ROTC Applicants	Air Force ROTC Graduates Receiving Commissions	Commissioned Air Force ROTC Graduates Selected for Pilot Training	Selection Rate
Male-all	219,887 (81.4)	30,280 (85.0)	9,239 (97.5)	30.5%
Female-all	50,081 (18.6)	5,354 (15.0)	237 (2.5)	4.4
White (all)	212,238 (78.6)	31,066 (87.2)	8,955 (94.5)	28.8
Minorities (all)	57,503 (21.3)	4,568 (12.8)	521 (5.5)	11.4
Black	32,798 (12.1)	2,645 (7.4)	186 (2.0)	7.0
Hispanic	12,647 (4.7)	771 (2.2)	172 (1.8)	22.3
Other minority	12,058 (4.5)	1,152 (3.2)	163 (1.7)	14.1
Ethnic status unknown	227 (< 0.1)			
Total	269,968 (100)	35,634 (100)	9,476 (100)	26.6

SOURCE: Carretta (in press) and information provided by the Air Force.

interest because they affect who will be available in this pool of potential airline
applicants. Any military practices that work against the acceptance of minorities
and women into flight training, for example, will have downstream consequences
for the airlines. Military selection procedures are also of interest because they are
thought to influence how selection is conducted by at least some air carriers
(Damos, 1995), despite the fact that the military is selecting individuals for pilot
training, whereas the airlines are selecting from among job candidates who are
already experienced pilots.

Our review of the literature on military pilot selection provides rather dramatic evidence that minorities and women may be more interested in flight
training than their numbers in the pilot ranks would suggest. Carretta (in press)
recently examined candidates for Air Force flight training who were applying on
the basis of their participation in the Reserve Officers' Training Corps (ROTC).
(Applicants who were graduates of the Air Force Academy were not included in
the study because they are not subjected to the same selection procedures.) For
the period 1981-1992, 19 percent of the individuals applying to participate in Air
Force ROTC were women; 21 percent were minority (Table 5-4). Of those
actually receiving military commissions upon graduation from Air Force ROTC

programs, 15 percent were women and 13 percent were minority. However, 30 percent of the white men and 22 percent of the Hispanics with commissions were accepted for training, but only 4 percent of women and 7 percent of blacks were accepted. Thus, white men constituted almost 98 percent of the group selected for pilot training; minorities, both men and women, constituted only 5.5 percent.

The Pilot Candidate Selection Method [5] used by the Air Force to select pilot candidates from ROTC clearly had differential impacts on the selection of individuals from different groups into pilot training. This does not necessarily imply that the various measures used are biased or that minorities and women were discriminated against in the selection process. Such a conclusion cannot be drawn without knowing something about the range and distribution of capabilities within the various applicant pools. As Carretta (in press:15) points out: "It is possible that well qualified females and ethnic minorities are less inclined to view the Air Force as an attractive career choice. Another possibility is that females and ethnic minorities are less likely to take courses or pursue leisure interests that might increase their performance on the [Air Force Officer Qualifying Test]." Research indicates that, in terms of predicting success in flight training, the tests and composites used by the Air Force are not biased against women or minorities (Carretta, in press:13).

The primary problem with the current military selection system is that its predictive validities are low. The highest prediction of performance in undergraduate pilot training using all available predictors is only .426 (accounting for only slightly more than 18 percent of the variance), which means that a large amount of the variance in flight training performance is not accounted for (Carretta and Ree, 1994). Traditional explanations for this low level of prediction are psychometric: (a) a dichotomous dependent variable (pass/fail in undergraduate pilot training); and (b) a severe restriction of range on the predictor variables due to dealing with an already highly selected group of pilot candidates. Low levels of predictive validity also leave open the possibility that performance is driven by social factors not being tapped.

There is also another problem: what is being predicted is success in training, not performance on the job. Predicting job performance, however, requires job analyses (which are scarce and even nonexistent for some specialties like pilots flying military transports rather than fighter planes) and then the development of valid, reliable measures of performance (Damos, 1996).

Since the military pays for pilot training, selecting candidates on the basis of procedures designed to predict training success has some justification. The situation is more complicated at the airlines, where candidates are already experienced pilots; performance on the job is of more interest at this stage than success

---

[5]The method includes the following components: medical/physical fitness, college performance, previous flying experience, the Air Force Officer Qualifying Test, and the Basic Aptitudes Test.

in training. In addition, the job of the airline pilot is changing in ways that raise uncertainties about whether the present selection procedures tap into the skills critical for success in piloting modern aircraft in a commercial setting.

Many current policies guiding pilot selection criteria were developed when aircraft were far less reliable than they are today and emphasized technical proficiency standards appropriate for single-pilot aircraft. Modern commercial aircraft, by contrast, are highly reliable, utilize computer systems management skills more than physical dexterity, and require pilots to work in a multicrew environment. As hardware failures have become less and less the cause of aircraft accidents, human errors have become more important, being implicated in two-thirds of all accidents worldwide (Foushee and Helmreich, 1988).

Human factors researchers (e.g., Damos, 1995, 1996; Helmreich et al., 1986) have highlighted the need for new job task analyses in modern aviation. They argue that pilot job task analysis needs to include attention to the social and personality skills associated with effective crew performance as well as the technical skills involved in flying the aircraft. From the employer side, Acton (1989) has described the central role that British Airways gave to a new job task analysis in reevaluating its selection criteria for ab initio pilot training. The company decided that six main pilot roles should serve as the basis for selection criteria: aircraft handler, systems manager, team leader, decision maker, communicator, and company representative. Similarly, Cathay Pacific, a Hong Kong-based carrier, used a new job analysis to guide its revision of selection procedures for choosing new first officers (copilots) and second officers (needed on so-called ultra long haul flights on the Boeing 747-400 airplane introduced in 1990). Cathay identified six areas of competence required of Cathay pilots: technical skill and aptitude; judgment and problem solving; communications (written and oral); social relations, personality, and compatibility with Cathay; leadership/subordinate style; and motivation and ambition (Bartram and Baxter, 1996:150-151).

Research in recent years has focused on predicting job performance, not just success in initial flight training. It has led human factors psychologists to identify a number of dimensions of personality factors that correlate with crew performance both in flight simulators and on the job as a commercial pilot (Chidester, Helmreich et al., 1990; Chidester, Kanki et al., 1990). Three profiles have been identified based on an algorithm developed by Chidester (1987, 1990). The most effective flight crew performance is associated with captains who exhibit both high achievement motivation and interpersonal skill. The least effective crew performance is associated with captains who are below average in achievement motivation and have a negative expressive style, such as complaining. An intermediate category includes captains who demonstrate high levels of competitiveness, verbal aggressiveness, impatience, and irritability. Crews may eventually adapt to this personal style, leading to adequate levels of performance.

Pilots fitting these various personality profiles can be identified through a self-report inventory, such as the Personal Characteristics Inventory (Pred et al., 1986; Spence and Helmreich, 1976; Spence et al., 1979). This inventory is reportedly being used in the pilot selection process by at least two commercial airlines. The Air Force is developing a test of skills required for effective crew coordination, communication, and decision making that will be added to its battery, drawing on similar concepts (Hedge et al., 1995).

The committee is not in a position to evaluate the research on pilot selection and performance or the new instruments being developed. The point of reporting the findings of our initial foray into selection procedures is to emphasize how much room there appears to be for new approaches. Concerns about the usefulness of existing pilot selection measures in predicting job performance in the modern commercial cockpit suggest that improvements are quite possible, at least in theory.

New approaches to pilot selection also offer the possibility of reducing selection barriers for women and minorities that are unrelated to their true ability to perform the job. Too little information is available in the public domain for us to judge the extent of experimentation with new selection approaches among the commercial airlines or the consequences of new selection measures on the characteristics of the pilot workforce or pilot performance on the job. All we can say at this point is that the experience of one airline (Northwest) that was willing to share its experience with us is promising.

Northwest recently began hiring pilots after a five-year abstention. The company has proclaimed its commitment to hiring the best employees at all levels, including the best and brightest women and minorities. It formed a task force to reevaluate its selection procedures from top to bottom. The airline continues to use flight time as one criteria in pilot selection but is giving increased attention to education, personality, and teamwork. It sees systems management skills as increasingly important as the company makes the transition to airplanes with glass cockpits. It is using a matrix approach to rating individuals, so that it can balance different selection factors within the overall pool of applicants. It has adopted the Personal Characteristics Inventory to screen for both positive and negative traits that are important to the company and has found that women and minorities perform better on this measure than on some traditional instruments. It is evaluating candidates on their ability to perform in a company with a diverse workforce. It continues to use a panel of company pilots for final selection (as most airlines do), but it has provided interview training to members of selection boards and has standardized interview questions. (Acton, 1989, also emphasizes the importance of training pilots on selection boards in interviewing techniques, as well as in the philosophy, process, and skills required to successfully identify good pilot candidates.)

More than 7,000 people applied for the 192 pilot jobs available at Northwest in 1995. Of the group selected, 5 percent were women and 34 percent were individuals the airline identifies as "people of color." This percentage of women exceeded the percentage of women with commercial and ATP licenses (4.2 percent and 2.5 percent, respectively). Although no data exist on the number of minorities with such licenses, the results at Northwest are clearly extraordinary based on employment levels in the industry as a whole.

The major airlines, because of their size, may be in the strongest position to initiate new approaches to pilot selection, but it is crucial that the smaller airlines pay attention to this issue as well. These airlines are increasingly likely to be the starting point for pilots beginning their airline flying careers. Their filtering role, as well as their training role, will grow in importance. The smaller airlines and the start-up airlines in fact may be ahead of their larger, older brethren, because they have been less dominated by military pilots and started operations at a time when more attention was being paid to the importance of a diverse workforce. Nevertheless, these individual companies may not be as well positioned as the majors to devote resources to reviewing the effects and effectiveness of their selection criteria and modifying them if necessary. One possible remedy is for the major airlines to work on this issue on behalf of their affiliated carriers; another is for the smaller airlines to address the issue jointly through their trade associations.

The committee recommends that all airlines examine their selection criteria and use procedures consistent with the best available knowledge of job tasks and effective crew performance.

School and Industry Climate

This chapter began with a recitation of the long struggle that minorities and women have faced to become full members of the aviation community. Throughout this report we have seen that victories have been achieved, but the battle is not yet completely won. The remaining job is in some ways even more difficult, because it requires addressing not blatant policies of discrimination and exclusion but habits of attitude and behavior that are much more difficult to identify and root out. It means creating a climate in which minorities and women can work as productively and comfortably as their white male colleagues and no invisible or artificial barriers restrict any individual's chances for advancement.

Aviation, like most of American society, isn't there yet, certainly not in the view of those who were excluded for so long. This can be seen in the frustration expressed by many blacks in congressional hearings in the mid-to-late 1980s examining discrimination in the airline industry, and by the objections raised by some groups not party to the case to ending the EEOC legal action against United in 1995, on the grounds that serious problems still remain. It can be seen in the work of Eiff et al. (1993), describing the conditions found during a firsthand look at maintenance technician training schools, conditions presenting "many subliminal and covert barriers to learning" for women and minorities that ranged from unequal facilities and expectations of students to the display of materials with offensive racial and sexual overtones. It can be seen in the incidents of harassment and hostility that have greeted some women and minority pilots when they "invaded" the formerly white male world of the cockpit (Henderson, 1995,1996; Wentworth, 1993).

Like much of American industry, aviation has to rid itself of some unfortunate legacies from its past. Also like other industries, aviation still has to shatter the so-called glass ceiling: the invisible, artificial barriers blocking women and minorities from advancing up the corporate ladder to management and executive positions (Federal Glass Ceiling Commission, 1995a:iii).[6] The Glass Ceiling report provides context for a discussion of the glass ceiling within the aviation industry.

The report found, in its comprehensive review of U.S. businesses, that the world at the top of the corporate hierarchy did not look anything like America, where two-thirds of the population, and 57 percent of the working population, is female, minority, or both (Federal Glass Ceiling Commission, 1995a:iii-iv). The report confirmed that at the highest levels of business there is a glass ceiling rarely penetrated by women or minorities. Specifically, it found (based on a 1989 survey) that 97 percent of the senior managers of *Fortune* 1,000 industrial and *Fortune* 500 companies were white and 95 to 97 percent were male; and that 3 to 5 percent of senior managers were women, and, of that percent, virtually all were white (Federal Glass Ceiling Commission, 1995a:12). Black men and women constituted less than 2.5 percent of total employment in the top jobs in the private sector (p. 9). We have no reason to assume that the airline industry represents a different profile from the profile of U.S. businesses generally or that the glass ceiling problems and issues described by the commission are not equally relevant and applicable to the airline industry.

As we have noted throughout this report, U.S. airlines have been peopled mostly by white men, and it is still the case that they dominate the management and piloting ranks of the industry. This may be changing, but not at a rapid pace—and the change did not start at a particularly early stage in the industry's growth (which, as we discuss, may have important implications for dealing with glass ceiling issues). For example, it has been suggested that as late as the 1970s, airlines did not have a consistent policy to encourage women who sought to work in management and executive positions (Douglas, 1991:96). In the 1980s, on the basis of its own examination of employment practices by three major air carriers, Congress found the industry wanting in providing employment opportunities for women and minorities in key professional and managerial positions (U.S. Con-

---

[6]The Glass Ceiling Commission, a 21-member bipartisan body appointed by President Bush and congressional leaders and chaired by the secretary of labor, was created by the Civil Rights Act of 1991. Its mandate was to identify the glass ceiling barriers that have blocked the advancement of minorities and women as well as the successful practices and policies that have led to the advancement of minority men and all women into decision-making positions in the private sector.

gress, 1988:23-4; 49), particularly noting the absence of minorities in the top tier of those companies (p. 9, 17, 19, 23). Factors that were found to have contributed to the lack of progress included the existence of a white male "old-boy" network for hiring and promotion decisions and the failure of airline companies to place minorities in mainstream positions or positions that are regarded as essential to company operations (U.S. Congress, 1988:3). A more recent assessment of women and minorities in the corporate offices and boardrooms of airline companies presents a more encouraging picture, with several impressive examples and illustrations of how the face of senior airline management is changing (Henderson, 1995, 1996).

A closer look at the Glass Ceiling report may help to bring into sharper focus the particular history and current status of women and minorities in senior airline management positions. The commission identified three levels of barriers to the advancement of women and minorities in businesses: societal, governmental, and internal structural barriers within the control of business (Federal Glass Ceiling Commission, 1995a:26-36). The internal structural barriers within the control of business include restrictive outreach and recruitment practices and pipeline barriers (initial placement and clustering in staff jobs that are not the career track to the top; lack of mentoring, management training, and opportunities for career development; and other internal practices that limit advancement) (pp. 32-36). The discussion of recruitment and pipeline barriers seems particularly relevant to the airline industry.

The Glass Ceiling research suggests that preparation for key corporate jobs requires 20 to 25 years "in the pipeline" with broad and varied experience in the "right" (core) areas of the business, which minorities and women generally have limited opportunity to obtain (pp. 15-16). The report notes that women and minorities tend to be in support, staff function areas and that movement between these positions and line positions is rare in most companies (pp. 15-16). We have already discussed that both women and minorities are not well represented in the core areas of the airline business—flight and maintenance operations and perhaps some others, such as sales, marketing, and finance. That has implications for both their recruitment and their movement within the airlines' corporate management structures. The linkage between experience in the core business of the airlines and success in moving up the corporate ladder has been recently highlighted (Henderson, 1996).

If it is necessary to be in the pipeline for 20 to 25 years and have access to and experience in the core business areas to make it into the top tier of corporate management, then women and minorities would seem to be facing a particularly difficult challenge to break into the airlines' corporate ranks in significant numbers in the near future. The numbers of women and minorities recruited and hired by the airlines, particularly as pilots, have been limited in the past. The airlines' interest and initiatives to more aggressively recruit, hire, and develop women and minorities for key management positions appear to have taken root later than has been the case for other major industries. It is unlikely, then, that there will be significant changes in the makeup of the industry's top leadership and management ranks any time soon. As the Glass Ceiling report points out, only businesses that sought to diversify their workforces in the late 1960s are now "cracking the glass ceiling," whereas most of those who started later are far behind (Federal Glass Ceiling Commission, 1995a:36). From every indication available to us, we would have to conclude that the airline industry falls in the latter category, with substantial progress yet to be demonstrated.

Drawing on the experience of businesses with successful programs and its own expertise and research, the Glass Ceiling Commission, in a separately published set of recommendations, has laid out a roadmap for businesses to follow in eliminating glass ceiling barriers (Federal Glass Ceiling Commission, 1995b:13-15):

• Demonstrate commitment by the chief executive officer by setting company-wide policies that actively promote diversity programs and policies that remove artificial barriers at every level.

• Include diversity in all strategic business plans and hold line managers accountable for progress.

• Use affirmative action as a tool, ensuring that all qualified individuals have equal access and opportunity to compete based on ability and merit.

• Select, promote, and retain qualified individuals, seeking candidates from noncustomary sources, background, and experiences and expanding the universe of qualified candidates.

• Prepare minorities and women for senior positions, expanding access to core areas of business and to various developmental experiences and establishing mentoring programs that provide career guidance and support to prepare minorities and women for senior positions.

• Educate the corporate ranks, providing formal training at regular intervals on company time to sensitize and familiarize all employees about the strengths and challenges of gender, racial, ethnic, and cultural differences.

• Initiate work-, life-, and family-friendly policies.

• Adopt high performance workplace practices.

We know from journalistic accounts (e.g., Henderson, 1996) and from information supplied by airlines to the committee that many airline companies now have in place or are developing policies and practices that track these broad recommendations. We are not in a position to evaluate these efforts, but we applaud and urge serious attention to them. We also acknowledge the reluctance of companies to share information on successful programs for competitive reasons but hope that the industry will try to find ways to learn what approaches are most successful at breaking down the barriers that hold back talented minorities and women.

The committee recommends continuing efforts, vigorously led by top officials, to root out any remaining vestiges of discriminatory behavior in aviation training institutions and aviation businesses and to provide a favorable climate and truly equal opportunities for all individuals who wish to pursue careers in the aviation industry.

References

CHAPTER 1
INTRODUCTION

Blue Ribbon Panel
1993   *Pilot and Maintenance Technicians for the Twenty-First Century—An Assessment of Availability and Quality.* Washington, DC: U.S. Department of Transportation, Federal Aviation Administration.
Douglas, Deborah G.
1991   *United States Women in Aviation, 1940-1985.* Washington, DC: Smithsonian Institution Press.
Federal Aviation Administration
no date   *FAA Statistical Handbook of Aviation: Calendar Year 1993.* FAA APO-95-5. Washington, DC: U.S. Department of Transportation.
Henderson, Danna K.
1995   The drive for diversity. *Air Transport World* (September):33-43.
National Commission to Ensure a Strong Competitive Airline Industry
1993   *Change, Challenge and Competition. A Report to the President and Congress.* Washington, DC: U.S. Government Printing Office.

CHAPTER 2
THE AVIATION INDUSTRY AND ITS WORKFORCE

Blue Ribbon Panel
1993   *Pilots and Aviation Maintenance Technicians for the Twenty-First Century: An Assessment of Availability and Quality.* Report of the Pilot and Aviation Maintenance Technician Blue Ribbon Panel. Washington, DC: U.S. Department of Transportation, Federal Aviation Administration, August.

Bureau of Labor Statistics
1985   *Industry Wage Survey: Certificated Air Carriers, June 1984.* August. Bulletin 2241. Washington, DC: U.S. Department of Labor.
1990   *Industry Wage Survey: Certificated Air Carriers, January 1989.* March. Bulletin 2356. Washington, DC: U.S. Department of Labor.
1994   *Employment and Earnings.* January, Vol. 41, No. 1. Washington, DC: U.S. Department of Labor.
1995a   *Employment and Earnings.* January, Vol. 42, No. 1. Washington, DC: U.S. Department of Labor.
1995b   *Occupational Compensation Survey: Pay and Benefits Certificated Air Carriers, March 1995.* September. Washington, DC: U.S. Department of Labor.

Bureau of the Census
1980   *1980 Census of Population: Classified Index of Industries and Occupations, First Edition.* October. Washington, DC: U.S. Department of Commerce.
1984   *1980 Census of Population: Volume 1, Characteristics of the Population, US Summary.* PC 80-1-D1-A. Washington, DC: U.S. Department of Commerce.
1992   *1990 Census of Population: Supplementary Reports, Detailed Occupation and Other Characteristics from the EEO File for the United States.* 1990 CP-S-1-1. Washington, DC: U.S. Department of Commerce.

Darby, Kit
1994   Slow rise in hiring but salaries stagnant. *Professional Pilot* June:44-46.

Davies, R.E.G.
1972   *Airlines of the United States Since 1914.* London: Putnam.

Equal Employment Opportunity Commission
1994   *Job Patterns for Minorities and Women in Private Industry 1993.* Washington, DC: EEOC.

FAPA
1993   *Pilot Employment Guide.* Atlanta, GA: FAPA.
1994   *Pilot Job Report.* Atlanta, GA: FAPA.
1995   *Job Market Update: Sorting Through the Facts & Fiction of the Industry.* Atlanta, GA: FAPA.

Federal Aviation Administration
n.d.(a) *Aviation Careers Series: Pilots and Flight Engineers.* PA-121-91. Washington, DC: U.S. Department of Transportation.
n.d.(b) *FAA Statistical Handbook of Aviation. Calendar Year 1993.* Washington, DC: U.S. Department of Transportation.
n.d.(c) Part 66-The New Certification Regulations for Aviation Maintenance Personnel. Draft (Jan. 5, 1996). U.S. Department of Transportation, Washington, DC.
1995   *FAA Aviation Forecasts: Fiscal Years 1995-2006.* Washington, DC: U.S. Department of Transportation.

Fitzgerald, Gerry
n.d.   *The Global Aviation Repair Shop: The Future of Overhaul and Maintenance.* Washington, DC: McGraw Hill Aviation Week Group.

Henderson, Danna K.
1995   The drive for diversity. *Air Transport World* September: 33-43.

Komons, Nick A.
1978   *Bonfires to Beacons: Federal Civil Aviation Policy Under the Air Commerce Act, 1926-1938.* Washington, DC: U.S. Government Printing Office.

Meyer, John R., and Clinton V. Oster, Jr., eds.
1981   *Airline Deregulation: The Early Experience.* Boston: Auburn House Publishing Company.

Morrison, Steven A., and Clifford Winston
1995   *The Evolution of the Airline Industry.* Washington, DC: Brookings Institution.
Nettey, Isaac Richmond
1995   Enhanced Integration of Multimodal Ground Transportation with Air Transportation at Selected Major Air Carrier Airports. Southwest Region University Transportation Center, Texas Southern University.
Petzinger, Thomas Jr.
1995   *Hard Landing: The Epic Contest for Power and Profits That Plunged the Airlines into Chaos.* New York: Times Books, a division of Random House.
Proctor, Paul
1995   Thriving regionals face labor unrest. *Aviation Week and Space Technology* 143(21):62.
Professional Pilot
1995   Propilot 1995 salary study. *Professional Pilot* 29(6):66-70.
Transportation Research Board
1991   *Winds of Change: Domestic Air Transport Since Deregulation.* Special Report 230. Washington, DC: National Research Council.
1996   *Future Aviation Activities: Ninth International Workshop.* Transportation Research Circular No. 454. Washington, DC: National Research Council.
U.S. Department of Transportation
1992   *Labor Relations and Labor Costs in the Airline Industry: Contemporary Issues.* DOT-P-30-92-1. Washington, DC: Office of the Secretary of Transportation.
Velocci, Anthony L., Jr.
1995   Unions pledge to take U.S. carriers to task. *Aviation Week and Space Technology* 143(21):73-76.
White, Andrea
1994   *Aviation Maintenance Careers.* Atlanta, GA: FAPA.
Wilbur Smith Associates
1995   *The Economic Impact of Civil Aviation on the U.S. Economy: Update '93.* Columbia, SC: Wilbur Smith Associates.
Wilson, Rosalyn A.
1994   *Transportation in America: Statistical Analysis of Transportation in the United States, 1994.* Landsdowne, VA: Eno Transportation Foundation, Inc.

CHAPTER 3
THE IMPACT OF MILITARY DOWNSIZING

Aspin, Les
1993   Policy on the assignment of women in the armed forces. Memorandum from the secretary of defense to the secretaries of the Army, Navy, and Air Force, et al., April 28, 1993. U.S. Department of Defense, Washington, DC.
1994   Direct ground combat definition and assignment rule. Memorandum from the secretary of defense to the secretaries of the Army, Navy, and Air Force, et al., January 13, 1994. U.S. Department of Defense, Washington, DC.
Binkin, Martin
1993   *Who Will Fight the Next War?* Washington, DC: Brookings Institution.
Binkin, Martin, and Shirley J. Bach
1977   *Women and the Military.* Washington, DC: Brookings Institution.
Binkin, Martin, and Martin J. Eitelberg with Alvin J. Shexnider and Marvin M. Smith
1982   *Blacks and the Military.* Washington, DC: Brookings Institution.

Blue Ribbon Panel
  1993   *Pilots and Aviation Maintenance Technicians for the Twenty-First Century: An Assessment of Availability and Quality.* Report of the Pilot and Maintenance Technician Blue Ribbon Panel. Washington, DC: U.S. Department of Transportation, Federal Aviation Administration, August.
FAPA
  n.d.   New-Hire Pilot Qualifications, 1985, 1987, 1988, 1990-1994. Atlanta, GA: FAPA.
  1994   *Aviation Maintenance Careers.* Atlanta, GA: FAPA.
Federal Aviation Administration
  1994   FAA-Air Force Agreement to Help Military Technicians Get Private Sector Jobs. Washington, DC: U.S. Department of Transportation. Internet document: FAA Office of Public Affairs Home Page.
Holm, Jeanne
  1992   *Women in the Military: An Unfinished Revolution,* rev. ed. Novato, CA: Presidio Press.
Kitfield, James
  1995   Preference and prejudice. *Government Executive* 27(6):23-29; 64.
Levy, Claire Mitchell
  1995   *The Civilian Airline Industry's Role in Military Pilot Retention: Beggarman or Thief?* Santa Monica, CA: RAND.
Thie, Harry J., William W. Taylor, Claire Mitchell Levy, Clifford M. Graf II, and Sheila Nataraj Kirby
  1994   *A Critical Assessment of Total Force Pilot Requirements, Management, and Training.* Santa Monica, CA: RAND.
  1995   *Total Force Pilot Requirements and Management: An Executive Summary.* Santa Monica, CA: RAND.
U.S. Department of Defense, Office of the Assistant Secretary of Defense, Personnel and Readiness
  1994   *Occupational Conversion Index—Enlisted/Officer/Civilian, September 1993.* DOD 1312.1. Washington, DC: U.S. Department of Defense.
U.S. President
  1995   Affirmative Action Review: Report to the President. Response to Presidential Directive, July 19, 1995. Washington, DC: White House.

CHAPTER 4
CIVILIAN TRAINING FOR AVIATION CAREERS

Aviation Week & Space Technology
  1989   Pilot turnover prompts regional airlines to expand, improve training programs. *Aviation Week & Space Technology* October 16:91-3.
Ayele, Moges
  1991   Attracting minorities to the transportation profession: Perspectives of historically black colleges and universities. *Transportation Research Record* 1305. National Research Council, Transportation Research Board, Washington, DC.
Blue Ribbon Panel
  1993   *Pilots and Maintenance Technicians for the Twenty-First Century—An Assessment of Availability and Quality.* Washington, DC: U.S. Department of Transportation, Federal Aviation Administration.
Damos, D.L.
  1996   Pilot selection batteries: Shortcomings and perspectives. *The International Journal of Aviation Psychology* 6(2):199-209.

FAPA
1992   *Pilot Employment Guide.* Atlanta, GA: FAPA.
Federal Aviation Administration
n.d.   *FAA Statistical Handbook of Aviation. Calendar Year 1993.* Washington, DC: U.S.
          Department of Transportation.
1990   *Airway Science Curriculum Demonstration Project: Qualitative Evaluation Report.* Pre-
          pared by Federal Aviation Administration, Higher Education and Advanced Technology
          Staff, AHT-30, July.
1994a   List of Certificated Pilot Schools, Advisory Circular No. 140-2W. July 28. Washington,
          DC: U.S. Department of Transportation.
1994b   Plan for Privatizing the Airway Science Curriculum Program. Report by the FAA Air-
          way Science Task Force. April 18.
1995   Part 147 Aviation Maintenance Technician Schools. National Vital Information System.
          March 2.
Fitzgerald, Gerry
n.d.   *The Global Aviation Repair Shop: The Future of Overhaul and Maintenance.* Washing-
          ton, DC: McGraw Hill Aviation Week Group.
Flight Training
1996   1996 flight school directory. *Flight Training* 8(5):53-76.
Garvey, William
1992   Airlines of the world train pilots with FlightSafety in US. *Professional Pilot* Febru-
          ary:68-70.
Glines, C.V.
1990   How will tomorrow's airline pilots be trained? *Air Line Pilot* September:18-21, 48.
Gore, Al
1993   *Creating a Government That Works Better and Costs Less: The Gore Report on Rein-
          venting Government.* New York: Times Books/Random House.
Griffin, Jeff
1990   *Becoming an Airline Pilot.* Blue Ridge Summit, PA: TAB Books (Division of McGraw-
          Hill, Inc.).
Horne, Thomas A.
1989   Lufthansa U: A visit to one of the world's most successful airline schools. *AOPA Pilot*
          32(5):79-85.
Hughes, David
1989   New approaches to pilot training stress human factors, coordination. *Aviation Week &
          Space Technology* October 16:86-7.
Kiteley, Gary W.
1995   Collegiate aviation programs and options. *Flight Training* 7(10):C45-9.
Morrison, Steven A., and Clifford Winston
1995   *The Evolution of the Airline Industry.* Washington, DC: Brookings Institution.
Morrocco, John D.
1995   Globalization drives British labor trends. *Aviation Week & Space Technology* Novem-
          ber 29:69-70.
National Center for Education Statistics
1994   *The Condition of Education, 1994.* NCES 94-149. Washington, DC: U.S. Department of
          Education.
Nelms, Douglas W.
1988   Training pilots from zero time to left seat. *Airline Executive* March:28-32.
1990   Training programs to ease pilot and mechanic labor shortages. *Airline Executive Inter-
          national* 14(2):16-19.

Oster, Clinton V., John S. Strong, and C. Kurt Zorn
1992   *Why Airplanes Crash.* Oxford: Oxford University Press.
Parke, Robert B.
1990   Where's the pilot shortage? *Business & Commercial Aviation* January:52-4.
Pisano, Dominick A.
1993   *To Fill the Skies with Pilots: The Civilian Pilot Training Program 1939-46.* Chicago: University of Illinois Press.
Proctor, Paul
1994   Regional carriers turn to flight schools. *Aviation Week & Space Technology* January 17:37.
Schukert, Michael A.
1994   *Post-Secondary Aviation & Space Education Reference Guide.* Washington, DC: U.S. Department of Transportation, Federal Aviation Administration.
Steenblik, Jan W.
1989   Pilot training for the next millennium. *Air Line Pilot* July:25-6.
Trent, William, and John Hill
1994   The contributions of historically black colleges and universities to the production of African American scientists and engineers. Pp. 68-80 in Willie Pearson, Jr., and Alan Fechter, eds., *Who Will Do Science? Educating the Next Generation.* Baltimore, MD: The Johns Hopkins University Press.
U.S. Congress, Senate
1989   Pilot supply and training. Hearing before the Subcommittee on Aviation, Committee on Commerce, Science, and Transportation, U.S. Senate. 101st Congress, First Session, August 3, S. Hrg. 101-307.
U.S. Department of Transportation
1992   *Labor Relations and Labor Costs in the Airline Industry: Contemporary Issues.* DOT-P-30-92-1. Washington, DC: Office of the Secretary of Transportation.
1993   *Report on the Audit of Federal Aviation Administration Sponsored Higher Education Programs.* Office of the Inspector General, Report #AV-FA-3-010. Washington, DC: U.S. Department of Transportation.
U.S. General Accounting Office
1994   *Management Reform: Implementation of the National Performance Review's Recommendations.* GAO/OCG-95-1. Washington, DC: U.S. General Accounting Office.
University Aviation Association
1990   The Impact of the Airway Science Program on Higher Education and Industry. Reprinted in FAA, Higher Education and Advanced Technology Staff, AHT-30, *Airway Science Curriculum Demonstration Project: Qualitative Evaluation Report.* July. Washington, DC: Federal Aviation Administration.
1994   *Collegiate Aviation Guide.* Auburn, Alabama: University Aviation Association.
Wallace, Joe
1989   From classrooms to cockpits. *Commuter Air* December:18-21.
Warwick, Graham, and J.W. Randall
1993   Ab initio training: Elementary introduction. *Flight International* 11-17 August:34-36.
White, Andrea
1994   *Aviation Maintenance Careers.* Atlanta, GA: FAPA.
Wilhelmsen, George R.
1995a   Airline prep school. *Plane & Pilot Aviation Careers* Special 1995 Report:30-1.
1995b   Fast track to the airline cockpit. *Plane & Pilot Aviation Careers* Special 1995 Report:23-5.
Wilkinson, Stephen
1991   The making of an airline pilot. *Air&Space* June/July:72-77.

CHAPTER 5
DIVERSIFYING THE AVIATION WORKFORCE

Acton, Robin
  1989   Selection of candidates for ab initio pilot training. Pp. 72-78 in Flight Safety Foundation, 42nd IASS, Athens, GA.
Bartram, Dave, and Peter Baxter
  1996   Validation of the Cathay Pacific Airways pilot selection program. *International Journal of Aviation Psychology* 6(2):149-169.
Board on Engineering Education
  1994   Major Issues in Engineering Education — A Working Paper. National Research Council, Washington, DC.
Brooks-Pasmany, Kathleen
  1991   *United States Women in Aviation 1919-1929*. Washington, DC: Smithsonian Institution Press.
Carretta, T.R.
  In press Group differences on U.S. Air Force pilot selection tests. *Military Psychology*.
Carretta, T.R., and J.J. Ree
  1994   Pilot-candidate selection method: Sources of validity. *The International Journal of Aviation Psychology* 4:103-117.
Chidester, T.R.
  1990   Trends and individual differences in response to short-haul flight operations. *Aviation, Space, and Environmental Medicine* 61:132-138.
  1987   Selection for optimal crew performance: Relative impact of selection and training. Pp. 473-479 in R.S. Jensen, ed., *Proceedings of the Fourth International Symposium on Aviation Psychology*, Columbus, OH.
Chidester, T.R., R.L. Helmreich, S.E. Gregorich, and C.E. Geis
  1990a   Pilot personality and crew coordination: Implications for training and selection. *The International Journal of Aviation Psychology* 1:23-42.
Chidester, T.R., B.G. Kanki, H.C. Foushee, C.L. Dickinson, and S.V. Bowles
  1990b   Personality Factors in Flight Operations: I. Leader Characteristics and Crew Performance in Full-Mission Air Transport Simulation. NASA Technical Memorandum No. 102259. NASA-Ames Research Center, Moffett Field, CA.
Clewell, Beatriz C., Bernice T. Anderson, and Margaret E. Thorpe
  1992   *Breaking the Barriers: Helping Female and Minority Students Succeed in Mathematics and Science*. San Francisco: Jossey-Bass.
College Board
  1995a   *Districtwide Reform: Lessons Learned from Equity 2000*. New York: College Board.
  1995b   *Trends in Student Aid: 1985-1995*. Washington, DC: College Board.
Contrearas, A., and O. Lee
  1990   Differential treatment of students by middle school science: Unintended cultural bias. *Science Education* 74:433-44.
Damos, D.L.
  1995   Pilot selection batteries: A critical examination. In *Proceedings of the 21st Conference of the European Association for Aviation Psychology. Volume 2: Aviation Psychology: Training and Selection*, N. Johnston, R. Fuller, and N. McDonald, eds. Brookfield, VT: Ashgate Publishing Company.
  1996   Pilot selection batteries: Shortcomings and perspectives. *The International Journal of Aviation Psychology* 6(2):199-209.
Douglas, Deborah G.
  1991   *United States Women in Aviation: 1940-1985*. Washington, DC: Smithsonian Institution Press.

Eiff, Gary M., Thomas K. Eismin, John Stahura, and Beverly A. Stitt
1993   Recruitment, Selection, Training and Placement of Non-Traditional Workers in Aviation
         Maintenance. Final Report. Submitted to GalaxyCorporation by Aviation Education
         Consultants, West Lafayette, IN, June 16.
FAPA
1993   *Pilot Employment Guide.* Atlanta, GA: FAPA.
Federal Glass Ceiling Commission
1995a   *Good for Business: Making Full Use of the Nation's Human Capital—The Environmen-
         tal Scan. A Fact Finding Report of the Federal Glass Ceiling Commission.* March 17.
         Washington, DC: U.S. Department of Labor.
1995b   *A Solid Investment: Making Full Use of the Nation's Human Capital. Recommenda-
         tions of the Glass Ceiling Commission. Final Report of the Glass Ceiling Commission.*
         November. Washington, DC: U.S. Department of Labor.
Foushee, H. Clayton, and Robert L. Helmreich
1988   Group interaction and flight crew performance. In *Human Factors in Aviation*, Earl L.
         Wiener and David C. Nagel, eds. Orlando, FL: Academic Press.
Griffin, Jeff
1990   *Becoming an Airline Pilot.* Blue Ridge Summit, PA: TAB Books (Division of McGraw-
         Hill, Inc.).
Hardesty, Von, and Dominick Pisano
1983   *Black Wings-The American Black in Aviation.* Washington, DC: Smithsonian Institu-
         tion Press.
Hedge, J.W., M.A. Hanson, F.M. Seim, K.T. Bruskiewicz, W.C. Borman, and K.K. Logan
1995   Development and validation of a crew resource management selection test for Air Force
         transport pilots. In *Proceedings of the Eighth International Symposium on Aviation
         Psychology*, R.S. Jensen and L.A. Rakovan, eds. Columbus, OH: Ohio State University.
Helmreich, R.L, L.L. Sawin, and A.L. Carsrud
1986   The honeymoon effect in job performance: Delayed predictive power of achievement
         motivation. *Journal of Applied Psychology* 71:1085-1088.
Henderson, Danna K.
1995   The drive for diversity, part 1. *Air Transport World* September:33-43.
1996   The drive for diversity, part 2. *Air Transport World* March:63-69.
Holm, Jeanne
1992   *Women in the Military—An Unfinished Revolution*, rev. ed. Novato, CA: Presidio Press.
Hunter, David R, and Eugene F. Burke
1995   *Handbook of Pilot Selection.* Aldershot, England: Avebury Aviation.
Institute of Medicine
1994   *Balancing the Scales of Opportunity: Ensuring Racial and Ethnic Diversity in the Health
         Professions.* Committee on Increasing Minority Participation in the Health Professions.
         Washington, DC: National Academy Press.
Jaynes, Gerald David, and Robin M. Williams, Jr., eds.
1989   *A Common Destiny: Blacks and American Society.* Committee on the Status of Black
         Americans. Washington, DC: National Academy Press.
Kahle, J.B., and M.K. Lakes
1983   The myth of equality in science classrooms. *Journal of Research in Science Teaching*
         20:131-40.
Kjos, Kristine
1993   A look at attitudes toward women in aviation—part 2. *FAA Aviation News.* April:21-24.
Malcom, S.M., and D.E. Chubin
1994   Structural Reform, Public Policy and Underserved Populations. Conference draft, pre-
         pared for CURIES Conference on Women in Science, Mathematics, and Engineering.
         Wellesley College, May 19-22.

Maples, Wallace R.
1992   *Opportunities in Aerospace Careers.* Lincolnwood, IL: VGM Career Horizons.

Mark, Robert
1994   *Becoming a Professional Pilot.* Blue Ridge Summit, PA: TAB Books (Division of McGraw-Hill, Inc.).

Marshall, Ray, and Marc Tucker
1992   *Thinking for a Living: Work, Skills, and the Future of the American Economy.* New York: BasicBooks.

Matyas, M.L., and S.M. Malcom, eds.
1991   *Investing in Human Potential: Science and Engineering at the Crossroads.* Washington, D.C.: American Association for the Advancement of Science.

Mitchell, Frank G.
1994   Success in aviation education: A national survey of secondary aviation magnet programs. *Journal of Aviation/Aerospace Education and Research* 5(1):5-7.

National Association of State Aviation Officials
1994   The States' Involvement in Aviation Education, July 1994.   A report prepared for the Federal Aviation Administration, Office of Training and Higher Education, Aviation Education Division.

National Center for Educational Statistics
1996   *Advanced Telecommunications in US Public Elementary and Secondary Schools, 1995.* Washington, DC: U.S. Department of Education.

National Center on Education and the Economy
1990   *America's Choice: High Skills or Low Wages!* The report of the Commission on the Skills of the American Workforce. Rochester, NY: National Center on Education and the Economy.

National Coalition for Aviation Education
n.d.   *A Guide to Aviation Education Resources.* Washington, DC: US Department of Transportation, Federal Aviation Administration.

National Research Council
1994   *NASA's Education Programs: Defining Goals, Assessing Outcomes.* Committee on NASA Education Program Outcomes. Washington, DC: National Academy Press.

National Science Board
1996   *Science & Engineering Indicators 1996.* NSB 96-21. Washington, DC: U.S. Government Printing Office.

National Science Foundation
1996   *Indicators of Science & Mathematics Education 1995.* NSF 96-52. Arlington, VA: National Science Foundation.

1994   *Women, Minorities, and Persons with Disabilities in Science and Engineering.* NSF 94-333. Arlington, VA: National Science Foundation.

1993   *Indicators of Science and Mathematics Education 1992.* NSF 93-95. Washington, DC: National Science Foundation.

National Science and Technology Council
1995   *A Strategic Planning Document for Meeting the 21st Century.* Washington, DC: National Science and Technology Council.

Nettey, I.R.
1996   African American pioneers in aviation. *History of Aviation: An American Experience,* T. Brady, ed. Carbondale, IL: Southern Illinois University Press.

Oakes, Claudia G.
1991a   *United States Women in Aviation Through World War I.* Washington, DC: Smithsonian Institution Press.

1991b   *United States Women in Aviation, 1930-1939*. Washington, DC: Smithsonian Institution Press.

Oakes, J.

1990a   *Lost Talent: The Underparticipation of Women, Minorities, and Disabled Persons in Science*. Santa Monica, CA: RAND.

1990b   *Multiplying Inequalities: The Effects of Race, Social Class, and Tracking on Opportunities to Learn Mathematics and Science*. Santa Monica, CA: RAND.

Office of Management and Budget

1995   *Budget of the United States Government, Fiscal Year 1996*. Washington, DC: U.S. Government Printing Office.

Office of Science and Technology Policy

1994   *Science in the National Interest*. Washington, DC: Office of Science and Technology Policy.

Oklahoma State Regents for Higher Education

1994   Systemwide Aviation/Aerospace Education Program Review. Aviation Aerospace Task Force's Report. January.

Pelavin, S.H., and M. Kane

1990   *Changing the Odds: Factors Increasing Access to College*. New York: College Entrance Examination Board.

Peng, S.S., D. Wright, and S. Hill

1994   *Understanding Racial-Ethnic Differences in Secondary School Science and Mathematics Education*. Washington, DC: U.S. Department of Education, National Center for Education Statistics.

Petzinger, Thomas, Jr.

1995   *Hard Landing*. New York: Random House.

Powell, William J.

1934   *Black Wings*. Los Angeles: Ivan Deach, Jr.

Pred, R.S., J.T. Spence, and R.L. Helmreich

1986   The development of new scales for the Jenkins Activity Survey measure of the Type A construct. *Social and Behavioral Science Documents*.

Quality Education for Minorities (QEM) Network

1990   *Laying a Foundation for Tomorrow: Report on the QEM Initial Years*. Washington, DC: QEM Network.

Rubovits, P.C., and Maehr, M.L.

1977   Pygmalion black and white. In *Contemporary Readings in Child Psychology*, E.M. Hetherington and R.D. Parkes, eds. New York: McGraw-Hill.

Shear, Michael D.

1995   They fought on two fronts—Tuskegee airmen recall war with Hitler—and Jim Crow. *Washington Post*. March 5: A1, A20.

Spence, J.T., and R.L. Helmreich

1976   *Masculinity and femininity: Their psychological dimensions, correlates, and antecedents*. Austin: University of Texas Press.

Spence, J.T. , R.L. Helmreich, and C.K. Holahan

1979   Negative and positive components of psychological masculinity and femininity and their relationship to self-reports of neurotic and acting out behaviors. *Journal of Personality and Social Psychology* 37:1673-1682.

Transportation Research Board

1992   *Civil Engineering Careers: Awareness, Retention, and Curriculum*. Report 347, National Cooperative Highway Research Program. Washington, DC: National Academy Press.

Trujillo, C.M.

1986   A comparative examination of classroom interactions between professors and minority
            and non-minority college students. *American Educational Research Journal* 23(94):629-
            642.

U.S. Congress, House of Representatives

1987a   *Discrimination Against Blacks in the Airline Industry.* Joint Hearing, Subcommittees of
            the Committee on Government Operations, September 30. Report No. 70-651. Wash-
            ington, DC: U.S. Congress.

1987b   *Trans World Airlines' Employment, Retention and Promotion of Blacks.* Hearing, Sub-
            committee of the Committee on Government Operations, December 3. Report No. 70-
            205. Washington, DC: U.S. Congress.

1987c   *Discrimination Against Blacks at United Airlines.* Hearing, Subcommittee of the Com-
            mittee on Government Operations, March 2. Hearings and Reports, Government Activi-
            ties and Transportation Subcommittee, Committee on Government Operations, Vol. 1,
            1987-1988. Washington, DC: U.S. Congress.

1988   *Slow Progress Regarding Affirmative Action in the Airline Industry.* Fifty-sixth Report
            by the Committee on Government Operations, July 19. Report No. 86-879. Washing-
            ton, DC: U.S. Congress.

U.S. Court of Appeals

1977   *EEOC v. United Airlines, Inc., et al.* 560 F. 2d 224 (June 28, 1977).

U.S. District Court

1976   Consent decree. *EEOC v. United Airlines, Inc., et al.* Northern District of Illinois,
            Eastern Division. Civil Action 73C 972 (April 30, 1976).

1995   Memorandum opinion and order. *EEOC v. United Airlines, Inc., et al.* Northern District
            of Illinois, Eastern Division. Civil Action 73 C972 (March 2, 1995).

U.S. General Accounting Office

1989   *Equal Employment Opportunity: Actions Needed for FAA to Implement Committee
            Recommendations in the Airline Industry.* GAO/HRD-89-100. Washington, DC: U.S.
            General Accounting Office.

1995   *Higher Education: Restructuring Student Aid Could Reduce Low-Income Student Drop-
            out Rate.* GAO/HEHS-95-48. Washington, DC: U.S. General Accounting Office.

U.S. Supreme Court

1963   *Colorado Anti-discrimination Commission v. Continental Airlines.* 372 US 714 (April
            22, 1963).

Vetter, Betty

1994   The next generation of scientists and engineers: Who's in the pipeline? Pp. 1-19 in
            Willie Pearson, Jr., and Alana Fechter, eds. *Who Will Do Science: Educating the Next
            Generation.* Baltimore, MD: Johns Hopkins University Press.

Wentworth, Kathleen

1993   Cockpit camaraderie or sexual harassment? *Air Line Pilot* January:17-20,55.

William T. Grant Foundation Commission on Work, Family and Citizenship

1988   *The Forgotten Half: Pathways to Success for America's Youth and Young Families.*
            Washington, DC: William T. Grant Foundation.

Appendix

Biographical Sketches

Clinton V. Oster, Jr. *(Chair),* is professor at the School of Public and Environmental Affairs at Indiana University. He served as research director for the Aviation Safety Commission in 1987-1988 and is former associate dean of the School of Public and Environmental Affairs and former director of Indiana University's Transportation Research Center. His current research focuses on airline economics, aviation safety, airport and airway infrastructure, and the environmental impacts of airline and airport operations. He served on the study committee for the Transportation Research Board (TRB) that in 1991 produced *Special Report 230: Winds of Change: Domestic Air Transport Since Deregulation* and chaired its Study Committee of the Federal Employers' Liability Act. He also served on the Office of Technology Assessment's Advisory Panel on Federal Aviation Research and Technology. He is active in the Transportation Research Forum, serving as president in 1995-1996. He has a B.S.E. from Princeton University, an M.S. from Carnegie Mellon University, and a Ph.D. from Harvard University.

Gary W. Baldwin is the research associate for the Committee on Education and Training for Civilian Aviation Careers. Previously, he was deputy director for labor and employee relations for the Federal Aviation Administration (FAA). He has held other senior management and staff positions with that and other federal agencies and has been a training consultant both to state and federal government agencies and to business organizations. He is an experienced labor negotiator who served as chief negotiator on labor agreements covering the FAA's air traffic controller and electronic technician workforces. He has a B.A. in political science from Seattle University and a J.D. from Gonzaga University School of Law.

Peggy Baty is the executive director of the International Women's Air and Space Museum in Dayton, Ohio. She is also the president and founder of Women in Aviation, International, an organization dedicated to promoting career opportunities in aviation for women. She has more than 16 years experience in aviation education, having served as dean of Parks College, associate vice chancellor and associate professor in Aeronautical Science at Embry-Riddle Aeronautical University, and department chair of aviation management at Georgia State University. She has a B.S. in aviation management and an M.Ed. in aerospace education from Middle Tennessee State University and a Ph.D. from the University of Tennessee. She is a commercial pilot and flight instructor, qualified in airplanes and helicopters.

Sandy Baum is a professor of economics and chair of the Department of Economics at Skidmore College. She has written widely on higher education finance and has worked with organizations such as the College Board, the National Association of Financial Aid Administrators, and the U.S. Department of Education on student financial aid issues. She has a B.A. from Bryn Mawr College and a Ph.D. from Columbia University.

Douglas C. Birdsall is vice president for marketing affiliates and strategic planning at Northwest Airlines. Prior to joining Northwest, he was president and chief executive officer of Travelmation Corporation and president and chief executive officer of New York Air and held senior management positions at Continental Airlines. He has extensive experience in airline management, marketing, planning, and finance. He has a B.S. in business administration from New York University and an A.A.S. in business administration from Westchester Community College; he also did graduate work at the Graduate School of Business Administration of New York University.

I.J. (Jim) Duncan became vice president for technical training for the Airbus Service Company's training center in 1991. He joined Aeroformation, the training arm of Airbus Industries, in Toulouse, France, in 1988. Prior to his assignment in Toulouse, where he served as a senior director of training, he was named to the management team at Airbus Training Center, also as senior director of training. He joined the airline industry in 1960 after active duty in the United States Air Force and served 24 years at Pan American World Airways in various flight crew and management positions, including vice president of flight operations. He retired from the Air Force in 1976, having served in the New York Air National Guard from 1964 until his retirement. He is a corporate trustee for the Council on Aviation Accreditation.

Jacqueline Fleming is president of the Motivation Research Corporation and an associate adjunct professor of psychology at Barnard College, Columbia University. She has specialized in motivation research and educational evaluation for the last 20 years. Much of this work has focused on the impact of college environments on African American students. She was formerly a consultant to the United Negro College Fund. She has a B.A. from Barnard College and a Ph.D. from Harvard University.

Janet S. Hansen is the study director for the Committee on Education and Training for Civilian Aviation Careers. As a senior program officer at the National Research Council, she has managed several projects related to education and training and to international comparative studies in education. Prior to joining the NRC staff, she was Director for Policy Analysis at the College Board. She wrote and lectured widely on issues relating to higher education finance, federal and state student assistance programs, and how families pay for college. She also served as Director for Continuing Education and Associate Provost at the Claremont Colleges and as Assistant Dean of the College at Princeton University. She graduated from the University of North Carolina and received a Ph.D. degree in public and international affairs from Princeton.

Captain M. Perry Jones has been in commercial aviation for more than 30 years. In 1965 he became the first African American pilot for Pan American World Airways and after 26 years joined Delta Airlines. During that time he has flown almost every type of commercial jet aircraft. He is active in a number of motivational programs for young people, including numerous mentoring programs, guest lecturing, and many appearances on television. Captain Jones received a B.S. in both mechanical and aeronautical engineering in 1959.

John Meyer is the James W. Harpel professor of capital formation and economic growth at Harvard University. In 1981-1983, he served as vice chairman of the Union Pacific Corporation. He also taught at Yale University from 1968 to 1973 and was president of the National Bureau of Economic Research, Inc., from 1967 to 1977. He serves as economic adviser or board member for several business firms and has been a consultant to the RAND Corporation, the World Bank, and the President's Council of Economic Advisers. He was a member of two presidential task forces on transportation and chaired a government commission studying railroad productivity. He also was a member of the Presidential Commission on Population Growth and the American Future. A 1958 Guggenheim Fellow, he received a B.A. from the University of Washington in 1950 and a Ph.D. in economics from Harvard University in 1955.

Isaac Richmond Nettey is the director of airway science at Texas Southern University. Prior to taking up his current position, he worked in airport operations at Houston Intercontinental Airport after teaching aviation at Northeast Louisiana University. He is treasurer of the University Aviation Association and chairman of the Airway Science Curriculum Committee. He has published work on intermodal transportation at large airports. He has received several awards in collegiate aviation education, including a certificate of commendation from the Federal Aviation Administration administrator in 1994 and a Partner in Education award from the Federal Aviation Administration in 1995. He was appointed an Eisenhower Transportation Faculty Fellow in April 1995. He serves on the FAA's Aviation Rulemaking Advisory Committee for airport certification, the board of directors of Sterling High School Aviation Sciences magnet program, and the Wings Over Houston Airshow. Nettey has a B.S. in aviation and an M.B.A. and was educated at Adisadel College, the University of Dubuque, Northeast Louisiana University, and the University of Houston.

Judith Orasanu is a principal investigator in the Flight Management and Human Factors Research Division of the NASA Ames Research Center at Moffett Field, California. Since 1989 she has conducted research on team problem solving, decision making, and communication in aviation. Prior to joining NASA, she managed basic research programs on cognition and training for the U.S. Army Research Institute and the U.S. Department of Education. She has a Ph.D. from Adelphi University in experimental psychology with an emphasis on psycholinguistics and human information processing. She received additional postdoctoral training in culture and cognition at the Rockefeller University.

Willie Pearson, Jr., is professor of sociology at Wake Forest University. He has served on numerous national panels, advisory boards, and committees, including the National Research Council's Committee on Women in Science and Engineering and the U.S. National Committee for the International Union of the History and Philosophy of Science. His primary research interest is in underrepresented groups in science and engineering. He has authored or coauthored numerous articles and books, including *Who Will Do Science? Educating the Next Generation.* He has a B.A. from Wiley College and a Ph.D. from Southern Illinois University at Carbondale.

Steven Sliwa is president of Embry-Riddle Aeronautical University, which has the largest U.S. collegiate programs dedicated to the aviation and aerospace industry. Prior to joining the university, he held several positions in aerospace and aviation-related fields as well as educational enterprises. He started his professional career with the National Aeronautics Space Administration (NASA), where he served in positions ranging from researcher to program manager to division level manager. He was founder and chief executive officer of an educational software firm that he operated for 8 years before selling to a larger software publisher; he was also founder and president of a nonprofit soaring training organization that is still in existence. He has a B.S.E. from Princeton University, an M.S.E. from George Washington University, and M.S. and Ph.D. degrees from Stanford University.

Harry J. Thie is a senior operations research and policy analyst with the RAND Corporation. Prior to that he served in a variety of military positions in the Department of the Army and the Department of Defense. He has conducted research and studies on military manpower, personnel, and training issues, including military pilot management and training. He has a B.A. from Saint Bonaventure University, an M.S. from the Georgia Institute of Technology, and a D.B.A. from George Washington University.

James C. Williams is general manager of the Engineering Materials Technology Laboratories at GE Aircraft Engines. He has a long-standing interest in technology policy as it affects various industrial sectors as well as the representation of women and minorities in science and engineering. Before joining GE he was professor of materials science and engineering and dean of engineering at Carnegie Mellon University. Prior to that he held research, engineering, and management positions with the Boeing Company and Rockwell International. He has B.S., M.S., and Ph.D. degrees in engineering from the University of Washington.

Fred Workley is the manager of maintenance operations for the National Air Transportation Association. He has 27 years experience in the aviation industry as a pilot, flight engineer, aviation maintenance technician, inspector, and instructor. He has airline, repair station, and association experience in maintenance and quality assurance on both large and small aircraft. He participates actively with the Federal Aviation Administration and other industry groups on the Aviation Rulemaking Advisory Committee and 19 working groups. He is a regular contributor of articles for the aviation press. Fred has an M.A. in management and quality control from the University of Phoenix.

Index

Index

job task analysis skills, 145
licensing requirements, 40-42, 79-80
military backgrounds, 4, 54-57
military downsizing, 59, 60-61, 69
military enlistment, trends in, 59-60,
69
minorities in military, 63, 69
minority and women, historical
development of, 115-119
non-airline, 40
numbers of, 2
on-the-job training, 5, 78-81
personality characteristics, 145-146
quality assessment, 95-96
selection criteria, 10-11, 42, 51, 142-
147
training pathways, 5-6, 77-78
wage trends, 36, 39-40
women in military, 65-69
worker supply and demand, 50-51,
105
workforce characteristics, 16, 30-31,
33
workforce diversity, 4, 44, 45-47
workforce projections, 99-103
See also Education and training for
aviation careers
Post Office, 19-20
Productivity, 22-23
pilot, 101-102
Public perception and understanding, 9
aviation image problems, 125-126
developing minority group interest in
aviation, 8-9, 119, 120-121
educational efforts for improving,
123-125
of industry hiring criteria, 10-11,
142
of technical/vocational education,
126-128

R

Regulatory environment
affirmative action oversight, 118-119
civil rights law, 117-118

commercial aviation classification,
25-27
competition, 20-21
current, 52
fare control, 22
historical development, 19-22
historical practice, 13, 16-17
labor costs and, 22, 36
Repairmen, 43, 44

S

Science education, 10
Seniority, 36, 39-40
Student loans, 139-141
Supplemental carriers, 21

T

Technicians
airframe and powerplant, 43, 81
certification, 42-44
collegiate training, 81, 83-84
current workforce, 16, 30-31
hiring patterns, 3, 50-51
military enlistment, historical
patterns of, 59-60
military training for, 57
minorities in military, 63, 69
numbers of, 2
on-the-job training, 5, 78, 81
school training for, 81, 83-84
training pathways, 5-6, 77-78
worker supply and demand, 50-51,
105
workforce diversity, 4
workforce projections, 102
See also Aviation maintenance
technicians; Mechanics
Technological developments, 23
airway science program training for,
107-108
basic academic competencies for,
128-129
comparative adaptability of training
pathways, 97